Frontiers in Applied Dynamical Systems: Reviews and Tutorials

Volume 2

W0230298

More information about this series at http://www.springer.com/series/13763

Frontiers in Applied Dynamical Systems: Reviews and Tutorials

The Frontiers in Applied Dynamical Systems (FIADS) covers emerging topics and significant developments in the field of applied dynamical systems. It is a collection of invited review articles by leading researchers in dynamical systems, their applications and related areas. Contributions in this series should be seen as a portal for a broad audience of researchers in dynamical systems at all levels and can serve as advanced teaching aids for graduate students. Each contribution provides an informal outline of a specific area, an interesting application, a recent technique, or a "how-to" for analytical methods and for computational algorithms, and a list of key references. All articles will be refereed.

Peter Jan van Leeuwen

Yuan Cheng • Sebastian Reich

Nonlinear Data Assimilation

Review 1: Peter Jan van Leeuwen: Nonlinear Data Assimilation for high-dimensional systems–with geophysical applications
Review 2: Yuan Cheng and Sebastian Reich: Assimilating Data into Scientific Models: An Optimal Coupling Perspective

 Springer

Peter Jan van Leeuwen
Department of Meterology
University of Reading
Reading, UK

Yuan Cheng
Intstitut fur Mathematik
University of Potsdam
Potsdam, Germany

Sebastian Reich
Intstitut fur Mathematik
University of Potsdam
Potsdam, Germany

ISSN 2364-4532   ISSN 2364-4931   (electronic)
Frontiers in Applied Dynamical Systems: Reviews and Tutorials
ISBN 978-3-319-18346-6   ISBN 978-3-319-18347-3   (eBook)
DOI 10.1007/978-3-319-18347-3

Library of Congress Control Number: 2015938920

Mathematics Subject Classification (2010): 86A22, 35R30, 93E11, 62M20, 65C05, 62F15

Springer Cham Heidelberg New York Dordrecht London

Printed on acid-free paper

Springer International Publishing AG Switzerland is part of Springer Science+Business Media (www.springer.com)

Preface to the Series

The subject of dynamical systems has matured over a period more than a century. It began with Poincaré's investigation into the motion of the celestial bodies, and he pioneered a new direction by looking at the equations of motion from a qualitative viewpoint. For different motivation, statistical physics was being developed and had led to the idea of ergodic motion. Together, these presaged an area that was to have significant impact on both pure and applied mathematics. This perspective of dynamical systems was refined and developed in the second half of the twentieth century and now provides a commonly accepted way of channeling mathematical ideas into applications. These applications now reach from biology and social behavior to optics and microphysics.

There is still a lot we do not understand and the mathematical area of dynamical systems remains vibrant. This is particularly true as researchers come to grips with spatially distributed systems and those affected by stochastic effects that interact with complex deterministic dynamics. Much of current progress is being driven by questions that come from the applications of dynamical systems. To truly appreciate and engage in this work then requires us to understand more than just the mathematical theory of the subject. But to invest the time it takes to learn a new sub-area of applied dynamics without a guide is often impossible. This is especially true if the reach of its novelty extends from new mathematical ideas to the motivating questions and issues of the domain science.

It was from this challenge facing us that the idea for the *Frontiers in Applied Dynamics* was born. Our hope is that through the editions of this series, both new and seasoned dynamicists will be able to get into the applied areas that are defining modern dynamical systems. Each article will expose an area of current interest and excitement, and provide a portal for learning and entering the area. Occasionally we will combine more than one paper in a volume if we see a related audience as we have done in the first few volumes. Any given paper may contain new ideas and results. But more importantly, the papers will provide a survey of recent activity and the necessary background to understand its significance, open questions and mathematical challenges.

Editors-in-Chief
Christopher K.R.T Jones, Björn Sandstede, Lai-Sang Young

Preface

Data assimilation is the science of combining information from prior knowledge in the form of a numerical dynamical model with new knowledge in the form of observations to obtain a best description of the system at hand. It is used to predict the future of the system, infer best parameter values, and to evaluate and compare different models. This 'best' description needs to contain information about the uncertainty, and the most general form is in terms of a probability distribution over the space of all possible model states. The basic mathematical formulation of the data assimilation problem is based on Bayes theorem, which states that this best probability distribution, called the posterior, is a point wise multiplication of the probability distribution of our prior knowledge from the numerical model with the probability distribution of the observations given each possible state of the model. The method is applied in almost all branches of science, although often under different names. Indeed, inverse modelling can be seen as a specific branch of data assimilation (or the other way around), as long as the model is dynamic in nature. Data assimilation can also often be formulated in terms of filtering and smoothing problems for stochastic processes.

An application of great practical relevance can be found in numerical weather forecasting, where atmospheric models and observations are combined every 6 to 12 hours to provide the best starting point for future forecasts. Other important applications can be found in all branches of the geosciences, such as oceanography, atmospheric pollution, marine biogeochemistry, ozone, seasonal forecasting, climate forecasting, sea-ice, glaciers and ice caps, ecology, land surface, etc. It is also used in oil reservoir modelling and seismology, and is typically referred to as history matching in those fields. Industrial applications are also widespread; think about all processes that need automatic control. Medical applications are growing too; data assimilation constitutes, for instance, an emerging field in the neurosciences.

Weather forecasting has been the driving force behind many recent theoretical and practical advances in data assimilation algorithms. The reason for this is twofold: the dimension of the system is huge, typically a billion nowadays, and the turnaround time is very short, the actual data assimilation can only take up to one hour, the rest of the 6- or 12- hour cycle is used to collect the billions of observations, perform quality control, which means throwing away close to 95% of the data, and preparing the observation-model interface. Currently used data-assimilation methods, which fulfil these operational constraints, can be divided into two categories: sequential methods and variational methods. This division is somewhat arbitrary, as will become clear shortly. The sequential methods are based on the Kalman filter. In the Kalman filter, the assumption is made that the probability distributions involved are all Gaussian. (In fact, the Kalman filter can also be derived assuming a linear update of the system, but that description falls outside the Bayesian framework.) The advantage of this approach is that only the first two moments of the distributions are needed. However, the size of the system in numerical weather prediction is too large to use the Kalman filter directly, simply because the second moments, the covariance matrix, need a billion squared entries. We have no supercomputer that can store that amount of numbers at present. Also, propagation of the covariance matrix under the model equations is prohibitively expensive. Perhaps, a bigger problem is that the Kalman filter is only justified for linear models. This limitation motivated the development of the ensemble Kalman filters starting in the 1990s in which the probability distribution is represented by a finite set of model states that are propagated with the full nonlinear model equations in between observations. Only at observation times, the ensemble of model states is assumed to represent a Gaussian and the Kalman filter update is implemented directly on the ensemble of model states. The Gaussian approximation can be justified from maximum entropy considerations given that only the ensemble mean and covariance matrix can be estimated from the available data. Furthermore, quite sophisticated methods have been developed to ensure efficient implementation, e.g. to avoid having to compute or store the full covariance matrix at any point in the algorithm. The finite ensemble size, typically 10–100 members can be afforded, leads to rank deficient matrices, and methods like localisation and inflation are used to counter this problem. These are to a large extent ad hoc, and this is a very active area of research.

The variational methods search for the mode of the posterior distribution. This can be the marginal posterior distribution at the time of the observations, leading to 3DVar, or the joint-in-time posterior distribution over a time window, in which case the method is called 4DVar. Again very sophisticated numerical techniques have been developed to solve this optimisation problem in billion-dimensional spaces. Unfortunately, these methods rely on linearisations and Gaussian assumptions on the prior and observation errors. Furthermore, the methods do not provide an uncertainty estimate, or only at very large cost.

Although numerical weather forecasting is quite successful, the introduction of convection resolving models and more complex observation networks leads to new challenges and the present-day methods will struggle. In particular, there is a strong need to move away from Gaussian data assimilation methods towards non-parametric methods, which are applicable to strongly nonlinear problems. (In numerical weather prediction, the so-called hybrids are becoming popular, combining ensemble Kalman filter and variational ideas, but this does not necessarily make the methods applicable to more nonlinear problems.) Fully non-parametric methods for probability distributions of arbitrary shape do exist and are based on sequential Monte Carlo methods, in which ensembles of model states, called samples, are generated to represent the posterior distribution. While these methods are extremely useful for small dimensional systems, they quickly suffer from the so-called curse of dimensionality, in which it is very unlikely for these states to end up in the high-probability areas of the posterior distribution.

Two solutions have been suggested to solve the curse-of-dimensionality problem. The first one is based on exploring the proposal density freedom in Monte Carlo methods. Instead of drawing samples from the prior distribution, one can draw samples from a proposed distribution and either accept them with a certain probability related to this proposal or change their weights relative to the other samples. This proposal density is chosen such that the samples will be from the high-probability area of the posterior distribution by construction.

Another option is to try to reduce the size of the problem by the so-called localisation. This reduces the influence of an observation to its direct neighbourhood, so that the actual data assimilation problem for each observation is of much smaller dimension. It is a standard method in ensemble Kalman filtering for high-dimensional systems, and it is key to their success.

This volume of Frontiers in Applied Dynamical Systems focuses on these two potential solutions to the nonlinear data assimilation problem for high-dimensional systems. Both contributions start from particle filters. A particle filter is a sequential Monte Carlo method in which the samples are called particles. It is a fully non-parametric method and applicable to strongly nonlinear systems. Particle filters have already found widespread applications ranging from speech recognition to robotics to, recently, the geosciences. The contribution of van Leeuwen focuses on the potential of proposal densities for efficiently implementing particle filters. It discusses the issues with present-day particle filters and explores new ideas for proposal densities to resolve them. A particle filter that works well in systems of any dimension is proposed and implemented for a high-dimensional example.

The contribution by Cheng and Reich discusses a unified framework for ensemble transform particle filters. This allows one to bridge successful ensemble Kalman filters with fully nonlinear particle filters and allows for a proper introduction of localisation in particle filters, which has been lacking up to now.

While both approaches introduce tuneable parameters into particle filters (such as the localisation radius), the proposed methods are capable of capturing strongly non-Gaussian behaviour in high dimensions. Both approaches are quite general and can be explored further in many different directions, making them both potential candidates for solving the full problem (for instance by combining them). We hope that they will form the basis of many new exciting ideas that push this field forward. As mentioned above, the number of application areas is huge, so high impact is guaranteed.

Reading, United Kingdom　　　　　　　　　　　　　　　Peter Jan van Leeuwen
Potsdam, Berlin, Germany　　　　　　　　　　　　　　　　　　　　Yuan Cheng
Potsdam, Berlin, Germany　　　　　　　　　　　　　　　　　　Sebastian Reich

Contents

Chapter 1
Nonlinear Data Assimilation for high-dimensional systems

- with geophysical applications -

Peter Jan van Leeuwen

Abstract In this chapter the state-of-the-art in data assimilation for high-dimensional highly nonlinear systems is reviewed, and recent developments are highlighted. This knowledge is available in terms of probability density functions, and the nonlinearity means that the shape of these functions is unknown a-priori. We focus on sampling methods because of they flexibility. Traditional Monte-Carlo methods like Metropolis-Hastings and its variants are discussed, including exciting new developments in this field. However, because of the serial nature of the sampling, and the possibility to reject samples these methods are not efficient in high-dimensional systems in which each sample is very expensive computationally. The emphasis of this chapter is on so-called particle filters as these are emerging as most efficient for these high-dimensional systems. Up to recently their profile has been low when the dimensions are high, or rather when the number of independent observations is high, because the area of state space when the observations are is decreasing very rapidly with system dimension. However, recent developments have beaten this curse of dimensionality, as will be demonstrated both theoretically and in high-dimensional examples. But it is also emphasized that much more needs to be done.

1 Introduction

1.1 What is data assimilation?

Loosely defined, data assimilation combines past knowledge of a system in the form of a numerical model with new information about that system in the form of observations of that system. It is a technique used every day in, e.g., numerical

P.J. van Leeuwen (✉)
Department of Meteorology, University of Reading, Earley Gate, Reading, UK

The National Centre for Earth Observation
e-mail: p.j.vanleeuwen@reading.ac.uk

© Springer International Publishing Switzerland 2015
P.J. van Leeuwen et al., *Nonlinear Data Assimilation*, Frontiers in Applied
Dynamical Systems: Reviews and Tutorials 2, DOI 10.1007/978-3-319-18347-3_1

weather forecasting, but its use is much more widespread. For instance, it is used in ocean forecasting, hydrology, space weather, traffic control, image processing, etc. Typically, the latter applications are considered inverse problems, and we discuss that view below.

This past knowledge is most generally represented by a probability measure. Since we will assume that the model of the system under study is finite dimensional we will assume this prior knowledge is available in the form of a probability density function, or pdf, $p(X = x)$, in which X is the random variable corresponding to model states, and $x \in \Re^{N_x}$ is a specific model state, represented as a state vector. This pdf is assumed to be known from model integrations.

The pdf denotes our uncertainty of the whereabouts of the true system, evolving according to

$$x_t^n = f_t(x_t^{n-1}) + \beta_t^n \tag{1.1}$$

in which the superscript is the time index, $f : \Re^{N_x} \to \Re^{N_x}$ maps the state one time step forward and denotes the deterministic part of the true evolution equation and β^n is the stochastic part of the true evolution. We have taken a discrete representation in time here. The dimension of the true system is typically much larger than of the modelled system. It could be infinite dimensional. Also, it is an interesting discussion whether the real system, so nature, is deterministic or stochastic. Some interpretations of quantum theory point to the latter, but the debate is still open. However, this discussion is not essential for our endeavour. The reason is that our (computer) model of the true system is always finite dimensional and will always contain errors, so our model system evolves like:

$$x^n = f(x^{n-1}) + \beta^n \tag{1.2}$$

in which the superscript is the time index, $f(x_t^{n-1})$ denotes that part of the true evolution equation that we have captured in our model, β^n is that part of the evolution equation that we have not modelled deterministically (e.g. because we don't know its exact form, or because we don't have the computer resources to take that part also into account, we come back to this later), represented here as a stochastic term. All we can ask for is the pdf of this model state.

The observations, represented by vector y, are taken from the real or true system under study and their uncertainty is also represented as a pdf: $p(Y = y|X = x_t)$, with $y \in \Re^{N_y}$ and in which x_t denotes the true system. This pdf is also assumed to be known, at least up to a normalisation constant. These observation vectors are finite dimensional as each actual measurement is a spatial and temporal average. Importantly, if we know this probability density as function of the state x_t, we know the probability density of this set of observations if they would arise from another state x, as $p(y|x)$, by just replacing x_t with x in the expression for $p(y|x_t)$. One issue here is of course that our model states are not at the same spatial and temporal resolution of the true system. This is one of the sources of the so-called representation errors (see, e.g., Cohn, 1997; Van Leeuwen, 2014), a proper discussion of which would be a paper on its own. Here we assume we know what $p(y|x)$ is, with x a state vector from our numerical model.

At the heart of data assimilation lies Bayes Theorem. The theorem is easily derived from the definition of a conditional probability density (pdf from now on) as follows. The conditional pdf of state x given a set of observations y, so $p(x|y)$, can be defined as:

$$p(x, y) = p(x|y)p(y) \tag{1.3}$$

Similarly we can define the pdf of the observations y given state vector x as:

$$p(x, y) = p(y|x)p(x) \tag{1.4}$$

Equating these two expressions for the joint pdf $p(x, y)$ we find:

$$p(x|y) = \frac{p(y|x)}{p(y)}p(x) \tag{1.5}$$

which is Bayes theorem. Although the derivation is simple, its interpretation is more subtle. Firstly $p(x)$ is the prior pdf which contains all our knowledge of the state vector before the observations are seen. Bayes Theorem then tells us how this pdf is updated by the observations y to form $p(x|y)$. What we need to do is multiply the prior pdf by $p(y|x)$, which is called the likelihood. To use this the observations have to be known, so y is a known vector, and $p(y|x)$ should be considered as a function of x. Finally, the result should be divided by $p(y)$, the marginal of the observations. We can write this function as:

$$p(y) = \int p(y, x)\, dx = \int p(y|x)p(x)\, dx \tag{1.6}$$

which shows that $p(y)$ is just a normalisation factor needed so that the posterior pdf integrates to one. The pdf $p(y)$ is NOT the pdf of the observations as such. It is the pdf of the observations before the actual measurement is made. This is typically hard to determine, so the expression in terms of $p(y|x)$ above gives a proper way to calculate it.

Bayes Theorem is a most powerful equation, and lies at the heart of all practical data-assimilation methods that have been developed (although it is sometimes difficult to trace it back to this because of the numerous implicit approximations made...).

1.2 How do inverse methods fit in?

As mentioned above, Bayes Theorem tells us that data assimilation is a multiplication problem: given the prior and the likelihood, the solution is the point-wise multiplication of the two. The argument that data assimilation is an inverse problem runs as follows. Reality is evolving in time, and at observation times we make observations of reality. Mathematically this can be expressed as:

$$y = H(\tilde{x}_t) + \epsilon \tag{1.7}$$

in which $\tilde{x}_t$ is the true state at model resolution, H is the operator that maps that state to observation space, and ϵ is a realisation of the measurement noise.

The argument now is that data assimilation is an inverse problem because we don't know the truth and we start from the observations y and try to recover $\tilde{x}_t^n$, or its evolution over time. The problem with this argument is that because the observations contain errors we know right from the start that this is impossible. There will always be a whole range of model states or model evolutions that could have generated the observations, typically an infinite number. This is true even in the case when we observe every dimension of the state vector. So the inverse problem formulated in this way is not satisfactorily thus far, but let us continue with this thinking as there are a few more issues with the approach.

Instead of the true solution one tries to find a 'best' solution. Mathematically this is formulated as a minimisation problem in which the true solution is approximated as the model state that is closest to all observations, e.g. in a least-squares sense or as minimum absolute distance. This part is quite often ad hoc. Why least squares? Only a coupling to the structure of the observation errors seems to make sense, and indeed, if the errors in the observations are Gaussian distributed a least squares approach will make sense.

An issue is that typically only part of the stat vector is observed, so even a least-squares approach will not lead to a unique solution. This is generally solved by introducing regularisation terms that are added to the least-squares penalty function. An example is a smoothness regularisation term that enforces, e.g., small derivatives. Or it is assumed that the solution should be close to another full solution. This can, for instance, be enforced again via an extra least-squares term (called ridge regularisation), or via an L1 term (the Lasso), or combinations of these. The literature on this subject is extensive and impressive, not least because of the extensive treatment of the numerical implementation, which is a big issue in itself. As an aside, one issue often overlooked with L1 regularisation (or any regularisation other than L2) is that the solution depends on the coordinate system used, which is a bit awkward. The regularisation term is typically chosen such that it leads to a well-defined unique solution of the inverse problem, both analytically and numerically. Or rather, the search is for a stable solution, meaning that a small change in the observations will lead to a small change in the solution. Given the fact that the observations do have errors and nature could have chosen another realisation of this observation error during the measurement process, looking for a stable solution does make sense.

However, the issue with the approach of adding regularisation terms is that they are ad hoc in the sense that they are not related to the physics of the problem but used to enforce a unique or stable solution. It then becomes unclear what the original problem was one is trying to solve, and different regularisations, not withstanding their sophistication, lead to different solutions. Also, it is good to realise that the search for the truth has been abandoned.

A more logical approach is to try to encapsulate all information one has on the solution in a prior before even looking at the observations. Since the prior knowledge is not exact we represent it by a probability density function (pdf).

That leads naturally to a Bayesian view of the data assimilation problem. This prior is then changed by the observations using Bayes Theorem, and the solution is an adapted prior called the posterior pdf. The problem is always well defined, ill-posedness does not appear at all, and neither does uniqueness. To me, this is a logical framework as is comes close to the simplest description of knowledge enhancement by us humans: we know our view of the world is not exact, and observations are used to sharpen that view. (One question is of course if we actually work with probabilities, probably not... And indeed, if a new observation comes up of something we had no prior of we tend to become quite upset, and often simply ignore that observation.)

Although intuitively perhaps more appealing, it must be mentioned that enormous difficulties arise when trying to apply Bayes Theorem to spatio-temporal random fields. The size of the problem is overwhelming, perhaps even prohibitively so. Think just about trying to store the prior pdf. Assume we have a 100 dimensional system, and want to use ten probability bins for each dimension to store the pdf in a computer. To do that we would need to store 10^{100} probabilities, which, given that the total number of atoms in the universe is estimated to be about 10^{80}, is a large number. So, necessarily, one needs efficient representations of the pdf (and efficient is not well defined). Also, how accurately can we perform the multiplications needed in Bayes Theorem? Viewed this way, there are ill-posedness problems within the Bayesian view too.

But let us set this discussion aside and concentrate on the real problem at hand, how to generate a good approximation of the posterior pdf as it appears in the geophysical data-assimilation problem.

1.3 Issues in geophysical systems and popular present-day data-assimilation methods

Geophysical systems are special in the sense that the observations are always sparse, observation errors can be quite large, the observation operator $H(x)$ can be strongly nonlinear, and the models used are high-dimensional (up to 10^9 for present-day numerical weather forecasting), and they can be highly nonlinear. Furthermore, error structures in observations and model are poorly known.

On the other hand, we do have model equations that describe the evolution of the system and that are often based on well-trusted physical conservation laws. Typically they are not exact conservation laws as there is always missing physics or biology or chemistry, either because we just don't know what a proper physical description is (e.g. fluid turbulence), or because the computer resources are too small to incorporate all we know. Reality is a combination of these two in most cases.

Given the size of these systems we might as well regard the data-assimilation problem as hopeless. The solution is defined in the introduction as the posterior pdf. Even if this pdf is Gaussian we still need to describe its first two moments, mean and covariance. For a state vector of size 10^9 the covariance has of the order of

10^{18} entries, more than any supercomputer in the world can store at this moment. Even when we realise that the majority of the entries might be very close to zero, long-range physical correlation does exist, so storing this matrix will at least fill up the largest supercomputer. (Note that as soon as computers become larger we will increase the resolution of the models used, so this problem will not go away.)

In real-world problems, however, the pdf is not Gaussian due to all nonlinearities involved. The pdf can have any shape which leads to even more problematic numbers as shown above.

So what can we do? Well, that is a difficult question, but we can study what has been done. At the moment two approaches have been followed. In the first it is assumed that the prior is Gaussian and the observations errors are Gaussian too, but the observation operator is allowed to be nonlinear. In this system one tries to find the most probable state of the system, the maximum of the posterior pdf, the (global) mode. This is found by looking for the minimum of $J(x) = -\log(p(x|y))$, which, because of the Gaussian assumptions mentioned above, leads to a sum of squares that has to be minimised. Note that the resulting so-called cost function, or penalty function, $J(x)$, is the same as the one appearing in the inverse literature as L2 regularisation or ridge regression, but now it is derived from first principles. This costfunction is minimised using gradient-descent techniques like conjugent gradient. The covariance matrix in the prior is stored in operator form, meaning that certain relations between model variables (called balances) are explored to reduce the actual size of the matrix to be stored. Although these relations are only approximately true, they do allow one to find meaningful solutions that, e.g., allow for skilful weather prediction. Unfortunately it is very difficult to find the posterior covariance, and this is typically not calculated in, e.g., numerical weather prediction. Another problem is that, since the optimisation is a local problem, we can only use the inverse of the local curvature, the Hessian, as the covariance. If $H(x)$ is a nonlinear function, the posterior is not a Gaussian, so the covariance that can be generated is only an approximation. We will discuss this method in more detail later.

Another popular method is the ensemble Kalman filter. This method represents the prior pdf by a number of model states, or ensemble members, typically of the order of 10–500 for geophysical applications. Observations are included by assuming that the prior is Gaussian, the observation errors are Gaussian and $H(x)$ is linear. In that case Bayes Theorem reduces to the Kalman filter, and different methods are employed to generate the posterior ensemble members, all using the Kalman filter equations in one form or the other. These equations are used even when $H(x)$ is nonlinear and the prior is actually non-Gaussian, so it is unclear what this method actually does in nonlinear systems (see, e.g., Le Gland et al., 2011). However, it is also used quite effectively in, e.g., numerical weather prediction. Also this method is discussed later in more detail.

The latest developments in numerical weather prediction are to combine the variational and ensemble Kalman filter approaches to find 'the best of both worlds'. Unfortunately the development is rather ad hoc, driven by operational needs. A more systematic approach that can handle fully nonlinear systems is desperately needed.

1.4 Potential nonlinear data-assimilation methods for geophysical systems

This paper is an attempt to show a possible way forward. As will be shown most standard nonlinear data-assimilation methods are not suitable for high-dimensional systems as they require an enormous amount of model runs, and each model run is a very expensive affair. However, strong developments have been reported recently that make these methods much more efficient, so they are included in the discussion as we need to keep an eye on their further development. In real geophysical applications we cannot afford more than 10 to perhaps 1000 model runs, which rules out most methods explored in other fields. The only exception to date are particle filters. As will be shown particle filters can handle fully non-linear systems and can provide more information than just the mode. Until recently common knowledge was that particle filters are impractical when the dimension of the model is larger than say 10, because the effective ensemble size quickly reduces to one or two, unless a very large number of particles is used. However, recent developments contradict this and efficient particle filters have been generated for high-dimensional systems in nonlinear settings. One of these methods claims even to be efficient independent of the size of the system at hand. A crucial ingredient in these efficient particle filters is their use of observations in a early state to generate particles close to the high probability peaks of the posterior pdf, and this is where the traditional techniques used in the geosciences like variational methods and ensemble Kalman filter ideas will come in handy.

1.5 Organisation of this paper

In this paper we will study nonlinear data assimilation for geophysical problems in a unified framework that will explore existing suboptimal data assimilation methods. The next chapter discusses standard nonlinear data-assimilation techniques based on Markov Chains, and these methods are contrasted with particle filters.

In Section 3 the basic idea behind particle filters is presented, followed by why this basic formulation can never work for large-dimensional systems. A general review on the application and usefulness of particle filters in geosciences is given in Van Leeuwen (2009), and a general overview of particle filtering is given by the excellent book by Doucet et al., 2001. There it was shown that although interesting progress had been made until 2009, no solution for the degeneracy problem, or the curse of dimensionality had been found. However, a lot of progress has been made the last couple of years, and these developments will be discussed here. We discuss resampling as a way to increase the efficiency of particle filters, and discuss why this is not enough for high-dimensional systems. Then we will discuss proposal densities, which form an important part of this paper. We show that they allow us an enormous amount of freedom to build particle filters for very high dimensional systems, and present an example of a successful approach that works in systems of any dimension by construction, illustrated by a high-dimensional example. This is the first example of undoubtedly an enormous growth in useful methods for extremely high dimensional systems encountered in the geosciences. The paper is closed by a comparison of several nonlinear data assimilation methods, both based on Metropolis-Hastings and several particle filters, on a simple system with variable dimension and a concluding section.

A final word of caution. Firstly, this survey is far from complete, although it tries to incorporate methods and ideas relevant for very high dimensional systems. Secondly, the paper does not strive for mathematical rigour, as will be clear from this introduction. Doing that would lengthen the paper considerably, at the expense of the overview. The philosophy employed here is that the ideas presented will encourage mathematicians to dive into this and help provide the necessary fundamentals.

2   Nonlinear data-assimilation methods

In this chapter a few popular nonlinear data-assimilation methods will be discussed to help understanding the issues involved in nonlinear data assimilation, and in particular, the application of these methods in the geosciences. One particular feature of the geosciences that stands out is the high dimension of the systems involved, easily 10^5–10^9, severely restricting the usefulness of several of the methods described here. We will briefly discuss the Gibbs sampler, several variants of Metropolis-Hastings, Hybrid Monte-Carlo, and Langevin sampling (MALA), and contrast them with particle filters. This is necessarily a very incomplete set, and one or more books can be written on all methods now available. Apart from particle filters, all of the methods discussed here have in common that they draw samples from a Markov Chain, so a new state vector trajectory is drawn based on the previous state-vector trajectory sample. While this is a logical strategy exploring previous good samples, it has as drawback that subsequent samples are dependent, leading to slow exploration of the state space. Indeed, as we will see, thousands of samples are typically needed for relatively small dimensional systems. Furthermore, extra samples are needed to obtain a good starting point of the Markov Chain, and most algorithms reject the majority of the samples generated to ensure convergence to the correct posterior pdf. In short, these Markov-Chain methods, even interesting new variants, are very inefficient. The method followed here is to present the basis ideas behind the methods rather than focusing on detailed proofs on concepts and convergence. More information on these methods can be found in, e.g., Robert and Casella (2004). A different methodology is the particle filter, which propagates independent samples by construction, so that much more efficient algorithms can be derived. That is why particle filters are the main subject of this paper, and expanded upon in the next chapters.

The standard setting is that we want to sample state vectors, or state-vector trajectories, that can be used as an efficient representation of the posterior pdf.

As discussed in the introduction we are interested in the best representation of the posterior pdf of the state (or trajectory) of our numerical model. This makes our data-assimilation problem discrete in space and time. Hence the state vector $x \in \Re^{N_x}$. Observations $y \in \Re^{N_y}$ are also assumed discrete in space and time.

2.1   The Gibbs sampler

The Gibbs sampler does not generate samples from the full posterior density directly but instead draws marginal samples in such a way that all these marginal draws together generate one new sample from the full density. Since each marginal is one- (or at least low-) dimensional, each of these draws is very cheap, making this a competitive scheme in certain circumstances. Let us see how it works in detail.

To sample from a complicated joint density, which is the posterior density in a data-assimilation problem, one can draw samples from the full marginals. Let us for simplicity write this joint density as $p(x^{(1)}, \ldots, x^{(N_x)}|y)$ in which the upper index is the spatial coordinate. The Gibbs sampler explores a Markov Chain to generate samples and works as follows:

1) Choose a first sample $x_0 = (x_0^{(1)}, x_0^{(2)}, \ldots, x_0^{(N_x)})^T$ from some initial density p_0.
2) Obtain a new sample x_n from x_{n-1} by sampling the values:

$$x_n^{(1)} \sim p(x_n^{(1)}|x_{n-1}^{(2)}, \ldots, x_{n-1}^{(N_x)}|y) \tag{1.8}$$

$$x_n^{(2)} \sim p(x_n^{(2)}|x_n^{(1)}, x_{n-1}^{(3)}, \ldots, x_{n-1}^{(N_x)}|y)$$

$$\cdots$$

$$x_n^{(d)} \sim p(x_n^{(d)}|x_n^{(1)}, \ldots, x_n^{(N_x-1)}|y)$$

3) Change n to $n + 1$ and proceed to 2) until convergence.

Note that each new component $x_n^{(i)}$ is used immediately to draw the next component $x_n^{(i+1)}$. 'Convergence' here means that we have reached the stationary joint distribution $p(x^{(1)}, \ldots, x^{(N_x)})$, and one sample is produced. Using properties of Markov Chains this convergence can easily be proved, see, e.g., Robert and Casella (2004). The number of samples needed to converge to the target pdf is called the burn-in period. Samples generated during this period cannot be used to infer the characteristics of the posterior pdf, making this method less attractive. As soon as we have generated this one sample we can either start again with a sample from p_0 or we can continue starting from this sample. Obviously, in the latter case the new sample we generate will be dependent on this sample, and we want to have independent samples. This means that the above procedure has to be performed several times to generate one next independent sample. How many times depends on the autocorrelation structure of this Markov Chain.

It is important to realise that the Gibbs sampler will only work if the marginal pdfs are easy to draw from. Any nonstandard pdf will make this method prohibitively expensive. As a general rule the Gibbs sampler can be used efficiently in two cases:

1) When the dimension of the system is small the conditional densities can be evaluated before hand, and sampling is just drawing from this small-dimensional density.
2) When the conditional densities are given in parametric form and samples can easily be generated from them.

If we now consider a large-scale geophysical application, e.g. numerical weather forecasting, the first is not the case, and the second is unclear.

As touched upon above, several sampling strategies are available to us, and typically one of the following is used:

1) Generate n chains using different starting points, each with m iterations to reach the stationary distribution. The costs to obtain n independent samples is mn.
2) Generate one single long chain and use the iterates after the m burn-in iterates to the stationary density. Sample only every kth value after the burn in period m. Costs for n independent samples $m + kn$ if k is large enough to ensure independent samples. Large enough is determined from the chain autocorrelation.
3) Combine 1) and 3), so l chains, typically $l < 10$, and keep each kth sample after m iterates. Cost for n independent samples is $lm + kn/l$, with k large enough, see 2).

Strategy 1) is not recommended because it is too expensive: if we want 100 independent samples and m is 10000 (which is a very low estimate), we would need 10^6 model runs, which is way too expensive indeed. If the posterior is relatively smooth without strong local maxima a single chain will do fine. Still, the cost is typically huge: taking $k = 10$, which is a rather low estimate again, we'd need about 20000 samples, still a lot. However, if the high-probability areas are almost disconnected, a very long chain is needed to jump to another high-probability area, so a single chain is ineffective. The third option with say $l = 5$ will lead to 52000 samples. It must be said that multiple chains allow for parallel implementation, which would significantly reduce the turn-around time. On the other hand, for a high-dimensional model of say 10^{6-9} model variables these numbers are all way too large.

Another way to increase the speed of convergence is the following. Instead of sampling one component of the state vector at a time one can sample a larger subset of the state vector. In high dimensional state spaces that is certainly recommended, especially when components are highly correlated. This is called block sampling.

In very high-dimensional spaces one typically encounters in the geosciences, one tries to generate a first sample directly on the stationary distribution to avoid the expensive and inefficient burn-in period. One of the possibilities is to first generate a 4DVar solution (a minimisation method that will be described in a later section) and start the Gibbs sampler from there. Note that this 4Dvar solution can be a local minimum. This has not been explored in any depth in the geosciences yet, but is potentially very interesting. Experience in numerical weather prediction shows that about 100 model runs (iterations) are needed to come close to the 4Dvar solution, which is much better than having to do the full burn in.

The following methods in this chapter all generate complete samples all at once, avoiding having to calculate the marginals first.

2.2 Metropolis-Hastings sampling

The Metropolis-Hastings sampler proposes a new sample, and then decides to accept it or not given some acceptance criterion. It works as follows:

1) Draw a starting point x_0 from some initial pdf p_0.
2) Move the chain to a new value z drawn from a proposal density $q(z|x_{n-1}, y)$
3) Evaluate the acceptance probability of the move $\alpha(x_{n-1}, z)$ with

$$\alpha(x_{n-1}, z) = min\left\{1, \frac{p(y|z)p(z)}{p(y|x_{n-1})p(x_{n-1})} \frac{q(x_{n-1}|z, y)}{q(z|x_{n-1}, y)}\right\} \tag{1.9}$$

4) Draw a random number u from $U(0, 1)$ and accept the move , i.e. $x_n = z$ when $u < \alpha$, otherwise reject it, so $x_n = x_{n-1}$.
5) Change n to $n + 1$ and return to 2) until convergence.

The **proposal density** is a very important density in Metropolis-Hastings, as it provides potential new samples. Most used proposal densities are the (multivariate) Gaussian and the Cauchy (or Lorentz) density centred around the current value of the chain, so x_{n-1} in this case. The covariance or width of these distributions determines the size of the step in state space and it typically adapted to the acceptance rate of the chain.

Obviously, this method is efficient if the acceptance rate is not too low. One way to achieve this is to make only small moves, so that α is close to 1, and hence the probability of acceptance is high. This would correspond to a small covariance or width of the proposal density. Unfortunately, this means that the chain moves very slowly through state space, and convergence will be slow. One somehow has to construct moves that are large enough to probe state space efficiently, while at the same time keep acceptance rates high. There are no general construction rules for the proposal q to do this. A practical solution is to monitor the acceptance rate, and adjust q such that acceptance rates are between 20% and 50%.

The remarks in the previous chapter about the sampling schemes, e.g. one single long or multiple shorter chains are valid here too. Also the convergence criteria carny over directly, so the method needs a very large number of model runs, especially since most model runs will not be accepted, leading to a valid new sample, but positioned on the previous sample.

We now discuss some specific proposal densities often used in Metropolis-Hastings algorithms.

1) If the proposal density q is symmetric, i.e. $q(x|z, y) = q(z|x, y)$ the acceptance rate α reduces to $\alpha = min\{1, p(y|z)p(z)/(p(y|x_{n-1}p(x_{n-1}))\}$, simplifying the calculation. This can be achieved, for instance, by choosing q a function of $|z-x|$.

2) An often used proposal density is that of the random walk, i.e. $z = x_{n-1} + \xi_n$, with ξ_n a random variable with distribution independent of the samples already obtained in the chain. Popular choices for q are Normal and Student's t. Clearly, the width of q determines the average size of the moves, and is typically chosen as a constant in the range $(0.5, 3)$ times the covariance of the chain.

3) The proposal density can also be chosen as not depending on the previous estimate x_{n-1}. (Note that the actual transition density $p(z|x)$ still does depend on x_{n-1}, so the chain is Markov.) A popular choice is the prior density, i.e. our best guess of the density before the new observations come into play. In this case, the acceptance ratio becomes the ratio of the likelihoods as Bayes Theorem shows. The main advantage is that α becomes relatively easy to calculate. However, when the prior and the likelihood disagree the prior samples will not be well positioned to describe the posterior.

4) Instead of updating the complete state vector at once, the so-called *global Metropolis-Hastings sampler*, one can also update individual components, or groups of components of the state vector, the *local Metropolis-Hastings sampler*.

5) Gibbs sampling can be seen as a special case of Metropolis-Hastings in which the proposal density is the density conditional of a certain component given the current values of the other components, as is easy to see.

6) Recent schemes use variants of the so-called preconditioned Crank-Nicolson algorithm, allowing for considerable speedup compared to the random walk method. It is described in detail below.

2.2.1 Crank-Nicolson Metropolis Hastings

Instead of using a simple random walk proposal, we can use a general Crank-Nicolson-like proposal as (see, e.g., Cotter et al., 2013 for an overview):

$$\left(I + \frac{1}{2}TB^{-1}\right)z = \left(I - \frac{1}{2}TB^{-1}\right)x_{n-1} + \sqrt{2T}\xi \tag{1.10}$$

in which B is the covariance of the prior, T is a preconditioning matrix to be chosen later, $\xi \sim N(0, I)$.

To understand the advantages of this kind of proposal let us look at the case $K = B$. This leads to the preconditioned Crank-Nicolson (pCN) proposal

$$z = \sqrt{1 - \beta^2}x_{n-1} + \beta\sqrt{B}\xi_n \tag{1.11}$$

in which $\beta = 2\sqrt{2}/3$. The acceptance ratio becomes

$$\alpha(x_{n-1}, z) = min\left\{1, \frac{p(y|z)p(z)}{p(y|x_{n-1})p(x_{n-1})}\frac{q(x_{n-1}|z)}{q(z|x_{n-1})}\right\} = min\left\{1, \frac{p(y|z)}{p(y|x_{n-1})}\right\} \tag{1.12}$$

The basic idea is to ensure that if no observations are present, every move is accepted. This leads to much better mixing of the algorithm, which is noteworthily of constant performance when the dimension of the system increases, in contrast with the standard random walk proposal. The reason for this performance is simply that the observations only inform about certain scales in the solution, and beyond those scales we rely on the prior. So if prior moves are always accepted beyond certain scales, as in the pCN, constant performance with scale reduction can be expected and is indeed found by Cotter et al. (2013). We thus find that this method will be more efficient than a simple random walk in high-dimensional systems.

It turns out that also without preconditioning, so choosing $T = I$ leads to acceptance ratio

$$\alpha(x_{n-1}, z) = min \left\{ 1, \frac{p(y|z)p(z)}{p(y|x_{n-1})p(x_{n-1})} \frac{q(x_{n-1}|z)}{q(z|x_{n-1})} \right\} = min \left\{ 1, \frac{p(y|z)}{p(y|x_{n-1})} \right\}$$
(1.13)

so again the prior is not part of the acceptance ratio, meaning, again, that scales smaller than those described by the data play no role in the acceptance criterion.

2.3   Hybrid Monte-Carlo Sampling

As mentioned before, if new candidates in the Metropolis-Hastings algorithm are chosen as in a random walk, as is usually done, the distance travelled through state space grows only as the square of the number of steps taken. Larger steps do not solve this as they lead to low acceptance rates.

Using ideas from dynamical systems we can make the method more efficient, see, e.g., Duane et al. (1987). The following is an introduction to dynamical systems, followed by the combination with Metropolis-Hasting to the hybrid scheme. The main difference with the standard Metropolis-Hastings scheme is that the hybrid method explores gradient information from the pdf, or more specifically $-\log p(x|y)$. In this sense it resembles variational methods to find the mode of the pdf, like 3DVar or 4DVar. Recall, however, that our goal with these sampling methods is not to find the mode, but to try to represent the full posterior pdf by a set of samples. The random ingredients in the method ensure that the methods do not just walk towards the mode.

2.3.1   Dynamical systems

We will exploit the evolution of a system under Hamiltonian dynamics to be defined shortly. Consider the evolution of state variable x under continuous time τ. In classical mechanics Newton's law describes how particles are accelerated by forces, the famous $F = ma$ in which m is the mass of the particle and a the acceleration, i.e. the second time derivative of the coordinates of a particle. This law being a second derivative in time, the evolution of a system of particles is fully determined if the forces are given, together with the position and velocities of the particles. So each particle is fully determined by its position and velocity coordinates, and these are related through $v^{(i)} = dx^{(i)}/dt$. The space spanned by the position and velocity variables is called phase space.

The probability density of the system can be written as:

$$p(x|y) = \frac{1}{Z_p} e^{-E(x)} \tag{1.14}$$

in which $E(x)$ is called the potential energy of the system when in state x. In a Hamiltonian system forces are conservative, which means that they can be written as the gradient of the potential energy, and Newton's law becomes:

$$\frac{dv}{d\tau} = -\frac{\partial E(x)}{\partial x} \tag{1.15}$$

The kinetic energy is defined as $K(v) = 1/2v^T v$, the sum of the squares of the velocities of the particles. The total energy of the system is the sum of the kinetic and potential energies:

$$H(x, v) = E(x) + K(v) \tag{1.16}$$

where H is called the Hamiltonian of the system. Because the velocities are the time derivatives of the positions and only the kinetic part of the Hamiltonian depends on the velocities, we find the Hamiltonian equations:

$$\frac{dx_i}{d\tau} = \frac{\partial H}{\partial v_i} = v_i \tag{1.17}$$

$$\frac{dv_i}{d\tau} = -\frac{\partial H}{\partial x_i}$$

During the evolution of the system the Hamiltonian is constant:

$$\frac{dH}{d\tau} = \sum_i \left\{ \frac{\partial H}{\partial x_i} \frac{dx_i}{d\tau} + \frac{\partial H}{\partial v_i} \frac{dv_i}{d\tau} \right\} \tag{1.18}$$

$$= \sum_i \left\{ \frac{\partial H}{\partial x_i} \frac{\partial H}{\partial v_i} - \frac{\partial H}{\partial v_i} \frac{\partial H}{\partial x_i} \right\} = 0$$

which is just energy conservation in this case. Note that this is related directly to the fact that the forces are assumed to be conservative.

Another important feature of Hamiltonian systems is that they preserve volume in phase space, the so-called Liouville Theorem. This can be derived by calculating the divergence of the flow field in phase space. This flow field is given by $V = (dx/dt, dv/dt)$ and its divergence is:

$$div V = \sum_i \left\{ \frac{\partial}{\partial x_i} \frac{dx_i}{d\tau} + \frac{\partial}{\partial v_i} \frac{dv_i}{d\tau} \right\} \tag{1.19}$$

$$= \sum_i \left\{ -\frac{\partial}{\partial x_i} \frac{\partial H}{\partial v_i} + \frac{\partial}{\partial v_i} \frac{\partial H}{\partial x_i} \right\} = 0$$

Having established these relations we can generate the laws that describe the evolution of the density of the system in a new way. Define the joint density of positions and velocities as:

$$p(x, v|y) = \frac{1}{Z_H} e^{-H(x,v)} \tag{1.20}$$

Because both H and the volume in phase space are conserved, the Hamiltonian dynamics will leave $p(x, v)$ invariant. The connection with sampling becomes apparent when we realise that although H is constant, x and v may change, and large changes in x are possible when v is large, avoiding random walk behaviour.

To exploit this we have to integrate the Hamiltonian equations numerically. Obviously, this will lead to numerical errors, that we would like to minimise. Especially, schemes that preserve Liouville's Theorem are of importance, as will become clear in the next section. One of those is the leap-frog scheme:

$$v(\tau + \epsilon/2) = v(\tau) - \frac{\epsilon}{2} \frac{\partial E}{\partial x}(x(\tau)) \tag{1.21}$$

$$x(\tau + \epsilon) = x(\tau) + \epsilon v(\tau + \epsilon/2)$$

$$v(\tau + \epsilon) = v(\tau + \epsilon/2) - \frac{\epsilon}{2} \frac{\partial E}{\partial x}(x(\tau + \epsilon))$$

Although each finite ϵ will lead to numerical errors, we will show in the next section that the hybrid scheme is able to compensate for them. But even with these numerical errors we see that numerical scheme follows Liouville's theorem because the first step changes all v by an amount that only depends on x. So along each line x is constant all v change the same amount, and this first step conserves volume. The same is true for the other two steps in the leap-frog scheme: the updated variable is changed by an amount that only depends on the other variable.

Finally, note that the main difference between the Metropolis-Hastings and the hybrid scheme is that the latter uses gradient information of the density (or rather the log density) while the former does not.

2.3.2   Hybrid Monte-Carlo

The hybrid scheme works as follows:

1) Add a random quantity to the velocity variables v by sampling from $p(v|x)$. Since that density is Gaussian it is relatively easy.
2) Update the position variable (which is the actual variable in the target density) with the leap-frog scheme, choosing a positive or negative value for ϵ each with probability $1/2$. Take L of those leap-frog steps.
3) Accept the new state (x^*, v^*) with probability

$$\alpha = min \left\{ 1, \exp \left[H(x, v) - H(x^*, v^*) \right] \right\} \tag{1.22}$$

4) Return to 1).

Note that one leap-frog step would make the scheme close to a random walk again, which is what we wanted to avoid. This is the reason to take L leap frog steps, with L not too small. Also note that the stochastic first step is needed to change the total energy of the system. If H would remain constant, the chain would not be ergodic.

An important question is how to choose the time step ϵ. Beskos et al. (2013) show that the time step in the leap-frog scheme should be of order $N_x^{1/4}$, leading to an acceptance rate of 0.651. This is related to a balance between the generating a proposal, which decreases as L increases (less time steps are needed to reach the set integration time), and the number of proposals needed to obtain acceptance, which increases as L increases.

If the numerical scheme used for solving the Hamiltonian equations was without error the value of H would not change, and each step will be accepted with probability one. Due to numerical errors, the value of H may sometimes decrease, leading to a bias in the Metropolis-Hastings scheme. To avoid this we need detailed balance. One way to achieve this is to choose the step in the leap-frog scheme random with equal probability for a positive or a negative value for ϵ, as we did in the scheme above.

In some applications a slight modification of the scheme is needed to avoid that the scheme returns to its initial position and ergodicity is lost. This can be avoided by choosing the step size in the leap-frog scheme random too. Finally, one could use a slightly different Hamiltonian in the leap-frog steps as long as the acceptance rate is determined using the correct Hamiltonian. This allows one to simplify the actual number of calculations needed.

In order to reduce the dependence in subsequent samples we can replace the Hamiltonian equations by (see Kim et al., 2003):

$$\frac{dx_i}{d\tau} = Av_i \qquad (1.23)$$

$$\frac{dv_i}{d\tau} = -A^T \frac{\partial H}{\partial x_i}$$

and try to choose matrix A such that the correlation between the subsequent samples is as small as possible. A disadvantage is that the samples are more expensive to generate. However, a useful property of a cleverly chosen matrix A is that matrix vector calculations can be performed efficiently in $N \log_2 N$ operations, leading to sometimes serious efficiency savings.

Also this method has not been applied to any real-sized geophysical problem, but we can make a judgement based on the number of operations needed to generate one sample. Firstly, one has to calculate $\partial E/\partial x$, which, for a nonlinear model especially, will not be trivial. Typically this will require numerical differentiation which is a complex task to code, and the model run needed to evaluate $\partial E/\partial x$ is typically more expensive than a standard forward run of the model. Secondly, the model has to be run twice for each leap-frog time step, for L leap frog time step. So if $L = 10$ we need the equivalent of 20 model runs to generate one new sample. So, again, to generate say 100 samples we need at least 2000 model runs.

A different approach is followed by Beskos et al. (2011), who study the formulation of Hybrid Monte Carlo on Hilbert spaces. Firstly they argue that when the acceptance rate only depends on the likelihood samples from the prior would always be accepted when the observations are worthless. This is useful because in that case the acceptance rate is robust under model refinement, as for the Crank-Nicholson proposals in Metropolis-Hastings. By formulating Hybrid Monte-Carlo on Hilbert spaces they enforce this requirement automatically. Then they notice that the covariance used for the velocity variables should have similar shape to that of the prior as large scales in the prior should allow for large velocities, so fast movement, in those directions where the length scale is large. The resulting Hamilton equations, and their numerical implementation via a modified leap-frog scheme outperforms some existing schemes.

It may be clear that a lot can be gained by studying more sophisticated splitting scheme of the Hamilton equations; the literature on this subject is growing fast, but is beyond the scope of this paper.

2.4 Langevin Monte-Carlo Sampling

If one uses the Hybrid Monte-Carlo algorithm with only one leap-frog step, Langevin Monte-Carlo sampling results. The behaviour of the system is close to that of the random walk.

Typically one omits the acceptance step by accepting all moves directly. In this case, there is no need to calculate the values of the new momentum variables v at the end of the leap-frog step since they will immediately be replaced by new values from the conditional density at the start of the next iteration. So, there is no reason to represent them at all. The scheme consists of the following steps:

1) Draw d random values $\beta^{(i)}$ from d Gaussian distributions $N(0, 1)$, where d is the dimension of the system.
2) Calculate a new value for the state vector via

$$x_n = x_{n-1} - \frac{\epsilon^2}{2} \frac{\partial E(x)}{\partial x} + \epsilon \beta \qquad (1.24)$$

which follows from contracting the momentum and position update in the leap-frog scheme.

One can show that if ϵ is small the acceptance rate in the Metropolis-Hastings version converges to one, so ignoring the acceptance step is justified.

Computationally the method needs only one evaluation of $\partial E/\partial x$ per sample generated. However, the subsequent samples will not be independent, so we will have to take several samples, so model evaluations, to generate one independent sample, with the number of samples needed dependent on the decorrelation time of the sample time series. In practice this would add a factor of at least 10 to the scheme, so, again, for independent samples one would need at least 1000 iterations of the scheme, not counting the burn-in period.

We can again make this algorithm much more efficient by exploring a Crank-Nicolson proposal (see Beskos et al., 2008) as follows:

$$\left(I + \frac{1}{2}TB^{-1}\right)z = \left(I - \frac{1}{2}TB^{-1}\right)x_{n-1} - TD\Phi(x_{n-1}) + \sqrt{2T}\xi \qquad (1.25)$$

in which $\Phi(x_{n-1}) = -\log p(y|x_{n-1})$, and again T is the preconditioning matrix. This method has been coined Metropolis-adapted Langevin (MALA).

Beskos et al. (2008) show that the acceptance ratio is only a function of the likelihood, so again superior scaling is expected for those cases where the model has finer resolution than the observations. This exciting result opens the possibility, again, for high-dimensional applications.

The conclusion from these extensions to the random walk Metropolis Hastings is that by choosing an appropriate Crank-Nicolson-based proposal perfect scaling of acceptance ratio with model resolution can be achieved. This opens ways to apply Metropolis-Hastings-like algorithms to high dimensional systems. However, it must be mentioned that perfect scaling does not necessarily mean that the methods are also efficient in high-dimensional applications. The number of samples needed for high-dimensional systems is still much larger than 10–100.

2.5   Discussion and preview

The disadvantage of all methods discussed so far is that by construction one has to perform several model runs to generate one new independent sample from the posterior. Furthermore, if no good first guess is available, one has to waste computer time on the burn-in period. The particle filter that we will discuss in the next sections doesn't have these problems, at least not by construction. If, however, a good first guess is available and the Markov chain mixes fast with reasonable acceptance rates, some methods do show potential. An example is Metropolis-Hastings with a preconditioned Crank-Nicolson proposal, see, e.g., Cotter et al. (2013). This is confirmed below when we compare several of these schemes with state-of-the-art particle filters.

On a practical level, since the samples are generated via a Markov Chain, it is not straightforward to make the algorithm parallel, unless more than one chain is used. Again, in a particle filter the model runs are typically independent from each other, so fully parallel. It is, in principle, possible to use 100 model runs for 100 independent samples from the posterior. As mentioned earlier, common knowledge on particle filters is that much more samples are needed (see, e.g., Snyder et al., 2008), but the rest of this article will try to prove that the field has advanced to making particle filters efficient for systems with arbitrary dimensions.

3   A simple Particle filter based on Importance Sampling

Particle filters are based on some form of Importance Sampling. A sample is drawn from a proposal density, often the prior, but instead of determining an acceptance ratio based on comparison with another sample, as in Metropolis-Hastings-like algorithms, each sample is weighted with the probability of that sample in the proposal compared to the posterior probability of that sample. This is done for several samples, after which all weights are compared to each other, and normalised. So, instead of accepting or rejecting a sample, each sample is accepted, and carries a weight. When, e.g., the sample mean is needed this is calculated as the weighted mean.

The above is the most straight-forward implementation of a particle filter and is called Basic Importance Sampling here. Basic Importance sampling is straightforward implementation of Bayes Theorem as we will show below. In the next chapters more sophisticated versions will be discussed.

3.1   Importance Sampling

The idea of importance sampling is as follows. Suppose one wants to draw samples from a pdf $p(x)$. The difficulty is that it is not easy to draw from this pdf, for instance because it is a combination of standard density functions. In importance sampling one generates draws from another pdf, the importance, or proposal pdf, from which it is easy to draw. To make these samples from the target pdf one reweighs the samples of the proposal by how probable that sample would be if drawn from the target pdf.

The idea is illustrated by the following example. Suppose we want to calculate the integral

$$I = \int f(x)p(x)\, dx \tag{1.26}$$

If we could draw samples from $p(x)$ directly, we would evaluate the integral as follows, using N samples $x_i \sim p(x)$:

$$I_N = \int f(x)\frac{1}{N}\sum_{i=1}^{N}\delta(x - x_i)\, dx = \frac{1}{N}\sum_{i=1}^{N}f(x_i) \tag{1.27}$$

However, as mentioned above, we cannot make these draws efficiently. So what we do is introduce a pdf $g(x)$ from which it is easy to draw, e.g. a Uniform or a Gaussian pdf. We can now write:

$$I = \int f(x)p(x)\, dx = \int f(x)\frac{p(x)}{g(x)}g(x)\, dx \tag{1.28}$$

and draw samples $x_i \sim g(x)$, to find:

$$I = \int f(x) \frac{p(x)}{g(x)} \frac{1}{N} \sum_{i=1}^{N} \delta(x - x_i) \, dx = \frac{1}{N} \sum_{i=1}^{N} f(x_i) \frac{p(x_i)}{g(x_i)} \tag{1.29}$$

So we can conclude that I am allowed to draw samples from a different density but I have to compensate for this by attaching an extra factor, or *weight* $p(x_i)/g(x_i)$ to each sample. In the following we explore this idea in Bayes theorem, using the posterior pdf as the target pdf and the prior pdf as the proposal pdf. In later chapter we will explore more exotic proposals to increase the efficiency of the method.

3.2 Basic Importance Sampling

The basic idea of particle filtering is to represent the prior pdf by a set of particles x_i, each of which is a full state vector or a full model trajectory as follows:

$$p(x) = \sum_{i=1}^{N} \frac{1}{N} \delta(x - x_i) \tag{1.30}$$

Using Bayes Theorem

$$p(x|y) = \frac{p(y|x)p(x)}{\int p(y|x)p(x) \, dx} \tag{1.31}$$

leads to

$$p(x|y) = \sum_{i=1}^{N} w_i \delta(x - x_i) \tag{1.32}$$

in which the weights w_i are given by:

$$w_i = \frac{p(y|x_i)}{\sum_{j=1}^{N} p(y|x_j)} \tag{1.33}$$

which is just a number we can calculate directly as detailed in the previous section.

Weighting the particles just means that their relative importance in the probability density changes. For instance, if we want to know the expectation of the function $f(x)$, we now have:

$$E[f(x)] = \int f(x)p(x) \, dx \approx \sum_{i=1}^{N} w_i f(x_i) \tag{1.34}$$

Common examples for $f(x)$ are x itself, giving the mean of the pdf, and the squared deviation from the mean, giving the covariance.

It is important to realise what we have done: we didn't draw samples from the posterior, as that is difficult, but we did draw samples from the prior. As mentioned in the first section in this chapter we should attach to each sample, or particle, a weight $p(x|y)/p(x)$. Because the posterior can be written as the normalised product of the prior and the likelihood, $p(x|y) = p(y|x)p(x)/p(y)$, only the normalised likelihood $p(y|x)/p(y)$ remains in the weights.

Let us now assume that the model is discrete in time and define $x^{0:n} = (x^0, \ldots, x^n)^T$ in which the superscript denotes the time index. A practical way to implement the particle filter is to calculate the one time state or the trajectory sequentially over time, which is where the name 'filter' comes from. The idea is to write the prior density as

$$p(x^{0:n}) = p(x^n|x^{0:n-1})p(x^{0:n-1}) \tag{1.35}$$

Using that the state vector evolution is Markov, i.e. all information needed to predict the future is contained in the present, and the past adds no extra information, we can write:

$$p(x^{0:n}) = p(x^n|x^{n-1})p(x^{0:n-1}) \tag{1.36}$$

Using this again on $p(x^{0:n-1})$, and on $p(x^{0:n-2})$, etc. we find:

$$p(x^{0:n}) = p(x^n|x^{n-1})p(x^{n-1}|x^{n-2})\ldots p(x^1|x^0)p(x^0) \tag{1.37}$$

Before we continue it is good to realise what the so-called transition densities $p(x^n|x^{n-1})$ actually mean. Consider a model evolution equation given by:

$$x^n = f(x^{n-1}) + \beta^n \tag{1.38}$$

in which $f(x^{n-1})$ is the deterministic part of the model and β^n is the stochastic part of the model. The stochastic part is sometimes called the model error, and the idea is that the model is not perfect, i.e. any numerical model used in the geosciences that is used to simulate the real world has errors (and these tend to be significant!). These errors are unknown (otherwise we would include them as deterministic terms in the equations) but we assume we are able to say something about their statistics, e.g. their mean, covariance, etc. Typically one assumes the errors in the model equations are Gaussian distributed with zero mean and known covariance, but that is not always the case. To draw from such a transition density $p(x^n|x^{n-1})$ means to draw β^n from its density and evaluate the model equation given above. In fact, for normal, or Gaussian, distributed model errors β^n with mean zero and covariance Q, we can write:

$$p(x^n|x^{n-1}) = N(f(x^{n-1}), Q) \tag{1.39}$$

stating that the mean of the pdf of x^n given the state at time $n-1$ is equal to the deterministic model evolution from that given state at time $n-1$, and the covariance of that pdf comes from the covariance of the stochastic part.

Model errors can be additive or multiplicative. Additive errors do not depend on the state of the system, while multiplicative errors do. We assume the model errors are additive in this paper to be able to derive analytical expressions later on and most progress has been made with this kind of errors.

Let us now continue with Importance Sampling. If we also assume that the observations at different times, conditional on the states at those times, are independent, which is not necessary for the formulation of the theory, but keeps the notation so much simpler, we have for the likelihood:

$$p(y^{1:n}|x^{0:n}) = p(y^n|x^n)\dots p(y^1|x^1) \tag{1.40}$$

where we used that y^j is not dependent on x^k with $j \neq k$ when x^j is known. The posterior density can now be written as:

$$p(x^{0:n}|y^{1:n}) = \frac{p(y^{1:n}|x^{0:n})p(x^{0:n})}{p(y^{1:n})} \tag{1.41}$$

$$= \frac{p(y^n|x^n)\dots p(y^1|x^1)p(x^n|x^{n-1})\dots p(x^1|x^0)p(x^0)}{p(y^n)\dots p(y^1)}$$

$$= \frac{p(y^n|x^n)p(x^n|x^{n-1})}{p(y^n)} \dots \frac{p(y^1|x^1)p(x^1|x^0)}{p(y^1)}p(x^0)$$

Realising that the last ratio in this equation is actually equal to $p(x^{0:1}|y^1)$ we find the following sequential relation:

$$p(x^{0:n}|y^{0:n}) = \frac{p(y^n|x^n)p(x^n|x^{n-1})}{p(y^n)}p(x^{0:n-1}|y^{1:n-1}) \tag{1.42}$$

This expression allows us to find the full posterior with the following sequential scheme (see Figure 1):

1 Sample N particles x_i from the initial model probability density $p(x^0)$, in which the superscript 0 denotes the time index.
2 Integrate all particles forward in time up to the measurement time. In probabilistic language we denote this as: sample from $p(x^j|x_i^{j-1})$ for each particle i, and each time j. Hence for each particle x_i run the model forward from time $j-1$ to time j using the nonlinear model equations. The stochastic part of the forward evolution is implemented by sampling from the density that describes the random forcing of the model.
3 Calculate the weights according to (1.33), normalise them so their sum is equal to 1, and attach these weights to each corresponding particle. Note that the particles are not modified, only their relative weight is changed!
4 Increase j by one and repeat 2 and 3 until all observations have been processed.

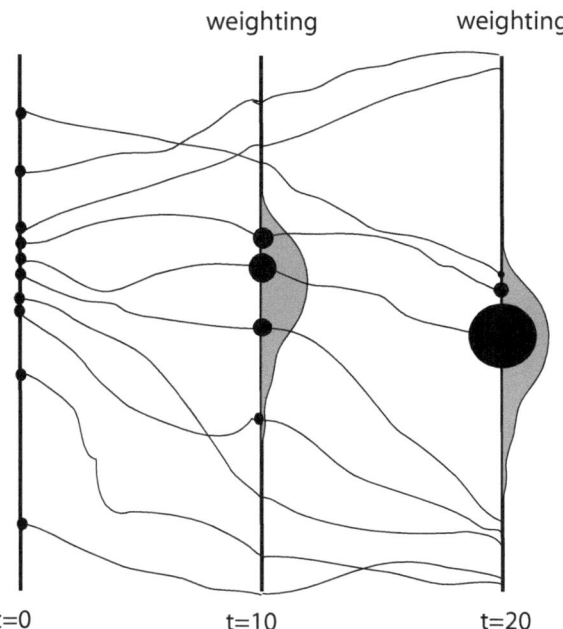

Fig. 1 *The standard particle filter with Importance Sampling. The model variable runs along the vertical axis, the weight of each particle corresponds to the size of the bullets on this axis. The horizontal axis denotes time, with observations at a time interval of 10 time units. All particles have equal weight at time 0. At time 10 the likelihood is displayed together with the new weights of each particle. At time 20 only 2 members have weights different from zero: the filter has become degenerate.*

It should be noted that the full posterior pdf over all time steps is created using a sequential algorithm. So, to find the marginal posterior pdf at time zero given all observations up to time n, $p(x^0|y^{1:n})$, one just has to use the full weights at time n over the whole trajectory of each particle. We thus see that the sequential filter algorithm gives rise to a smoother estimate.

The good thing about importance sampling is that the particles are not modified, so that dynamical balances present in a pure model integration (see introduction) are not destroyed by the data assimilation. The bad thing about importance sampling is that the particles are not modified, so that when all particles move away from the observations they are not pulled back to the observations. Only their relative weights are changed.

It is stressed how simple this scheme is compared to traditional methods like 3- or 4DVar and (Ensemble) Kalman filters. (These schemes are discussed later for those readers not familiar with them.) The success of these schemes depends heavily on the accuracy and error covariances of the model state vector. In 3- and 4DVar this leads to complicated covariance structures to ensure balances, etc. In Ensemble

Kalman filters artificial tricks like covariance inflation and localisation are needed to get good results in high dimensional systems. Particle filters do not have these difficulties because the covariance of the state vector is never used.

However, there is (of course) a drawback. Even if the particles manage to follow the observations in time, the weights will differ more and more. Application to even very low-dimensional systems shows that after a few analysis steps one particle gets all the weight, while all other particles have very low weights (see Figure 1, at $t = 20$). So the variance in the weights gets very large. That means that the statistical information in the ensemble becomes too low to be meaningful. For instance, if we would calculate the mean, so the weighted mean in this case, this mean is determined by the one particle with weight close to one and the others have no influence because their weight is so low. Furthermore, the weighted sample variance would be (very close to) zero. Clearly, this results in a very poor representation of the posterior pdf. This is called *filter degeneracy* or *ensemble collapse*. It has given importance sampling a low profile until resampling was invented, see the next section.

4 Reducing the variance in the weights

In the previous chapter we have seen how simple a particle filter with basic importance sampling is. However, we also noted that this filter will be degenerate for even very small-dimensional systems simply because we multiply weight upon weight. Several methods exist to counteract this behaviour, so to reduce the variance in the weights, and we discuss resampling and the Auxiliary Particle Filter here (see, e.g., Van Leeuwen, 2009, for other methods).

4.1 Resampling

In resampling the posterior ensemble is resampled so that the weights become more equal (Gordon et al., 1993). The idea of resampling is simply that particles with very low weights are abandoned, while multiple copies of particles with high weight are kept. In order to restore the total number of particles N, identical copies of high-weight particles are formed. The higher the weight of a particle the more copies are generated, such that the total number of particles becomes N again. Sequential Importance Re-sampling (SIR) does the above, and makes sure that the weights of all posterior particles are equal again, to $1/N$.

Sequential Importance Re-sampling is identical to Basic Importance Sampling discussed in the previous chapter but for a resampling step after the calculation of the weights. The 'flow chart' reads (see Figure 2):

1 Sample N particles x_i from the initial model probability density $p(x^0)$.
2 Integrate all particles forward in time up to the measurement time (so, sample from $p(x^n|x_i^{n-1})$ for each i)

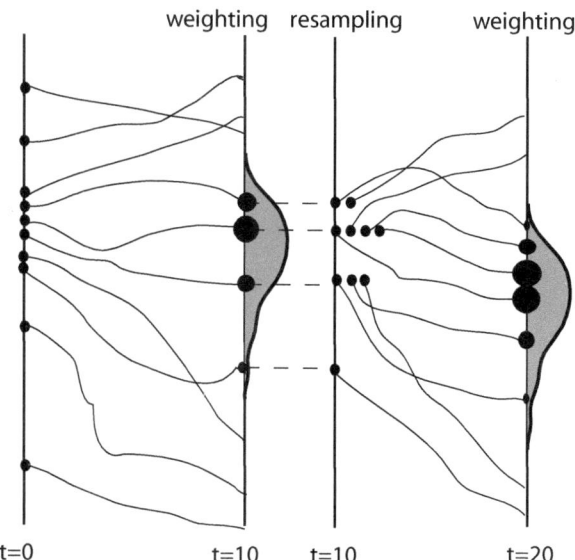

Fig. 2 *The Particle Filter with Resampling, also called Sequential Importance Resampling. The model variable runs along the vertical axis, the weight of each particle corresponds to the size of the bullets on this axis. The horizontal axis denotes time, with observations at a time interval of 10 time units. All particles have equal weight at time zero. At time 10 the particles are weighted according to the likelihood, and resampled to obtain an equal-weight ensemble.*

3 Calculate the weights according to (1.33) and attach these weights to each corresponding particle.

4 Re-sample the particles such that the weights are equal to $1/N$.

5 Repeat 2, 3 and 4 sequentially until all observations have been processed.

It is good to realise that the resampling step destroys the smoother character of the method. All particles that are not chosen in the resampling scheme are lost, and their evolution is broken. So the smoother estimate is built up of lesser and lesser particles over time, until it consists of only one particle, losing again all statistical meaning. However, as a filter the method is still very useful as at any time we have a full ensemble of particles given all previous observations.

The resampling can be performed in many ways, and we discuss the most used.

1) *Probabilistic resampling*

Most straightforwardly is to directly sample randomly from the density given by the weights. Since this density is discrete and one-dimensional this is an easy task. However, due to the random character of the sampling, so-called sampling noise is introduced. This method has been named generalised Bernoulli for those versed in the sampling literature.

2) *Residual Sampling*

In this re-sampling method all weights are multiplied with the ensemble size N. Then n copies are taken of each particle i in which n is the integer part of Nw_i. After obtaining these copies of all members with $Nw_i \geq 1$, the integer parts of Nw_i are subtracted from Nw_i. The rest of the particles needed to obtain ensemble size N are then drawn randomly from this resulting distribution.

3) *Stochastic Universal Sampling*

In this method all weights are put after each other on the unit interval $[0, 1]$. Then a random number is drawn from a uniform density on $[0, 1/N]$, and $N - 1$ line pieces starting from the random number, and with interval length $1/N$ are laid on the line $[0, 1]$. A particle is chosen when one of the end points of these line pieces falls in the weight bin of that particle. Clearly, particles with high weights span an interval larger than $1/N$ and will be chosen a number of times, while small weight particles have a negligible change of being chosen. While Residual Sampling reduces the sampling noise, it can be shown that Stochastic Universal Sampling has lowest sampling noise.

Snyder et al. (2008) prove that resampling will not be enough to avoid filter collapse and argue it is related to the dimension of the state vector. The following argument shows a slightly different interpretation. Let us consider two—artificial— particles, one always 0.1σ away from a set of N_y independent observations, and the other 0.2σ away from them, in which σ is the standard deviation of each observations, assumed to be equal for all N_y Gaussian distributed observations. This is admittedly not something that would happen in reality but it does help in bringing two important points home. The point now is that these are two almost perfect particles, what more can you want? The weight of the first particle will be:

$$w_1 = A \exp\left[-\frac{1}{2}(y - H(x_1))^T R^{-1}(y - H(x_1))\right] = A \exp(-0.005N_y) \qquad (1.43)$$

while the weight of the second particle is

$$w_2 = A \exp\left[-\frac{1}{2}(y - H(x_2))^T R^{-1}(y - H(x_2))\right] = A \exp(-0.02N_y) \qquad (1.44)$$

The ratio of these two weights is

$$\frac{w_2}{w_1} = \exp(-0.015N_y) \qquad (1.45)$$

Having a geophysical application in mind, let's assume we have a modest $N_y = 1000$ independent observations. This leads to a ratio of the weights of

$$\frac{w_2}{w_1} = \exp(-15) \approx 3 \ 10^{-7} \qquad (1.46)$$

so particle two has a negligible weight compared to particle 1 while both are absolutely excellent particles. We can draw two conclusions from this: 1) the number of independent observations is crucial, and 2) the accuracy of the observations is not the main factor, even inaccurate observations will give rise to filter degeneracy if the number of independent observations is large.

This argument is strengthened by the following. Consider the hypersphere with radius r around the observations in N_y dimensional space. The volume of that hypersphere is given by (see, e.g., Huber, 1982):

$$V \propto \frac{r_y^N}{\Gamma(N_y/2 - 1)} \tag{1.47}$$

Let's take for the radius of this hypersphere 3σ we find for this volume, using Stirling for the factorial:

$$V \propto \left[\frac{9\sigma}{N_y/2}\right]^{N_y/2} \tag{1.48}$$

This is a very quickly decreasing function of N_y as Figure 3 shows. So it is highly unlikely for particles to end up in this hypersphere, so closer than 3σ from the observations, when N_y is large. Again we see that the number of observations is the crucial part of degeneracy.

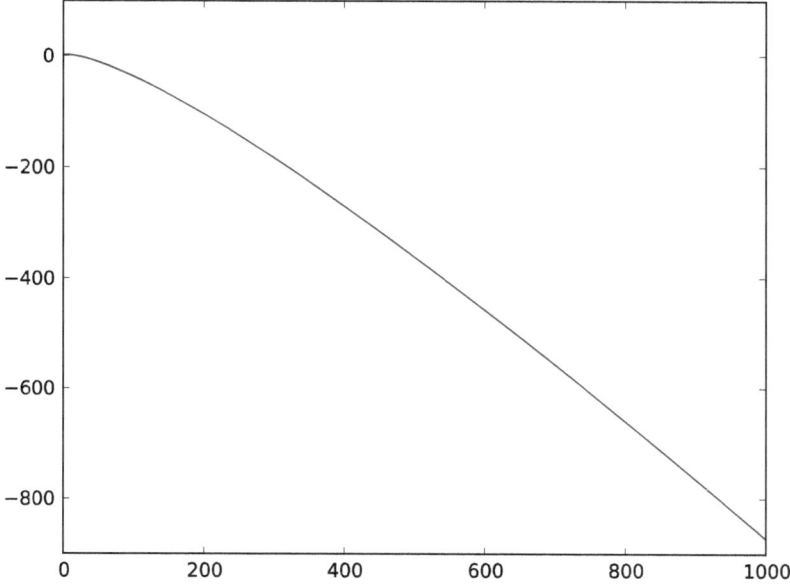

Fig. 3 $\log_{10}$ *of the volume in observation space of ball with radius 3 standard deviations (taken as 1 here) against dimension. Note the very rapid decrease of this function.*

So a reinterpretation of the Snyder et al. (2008) results, backed up in a later section when we discuss the optimal proposal density, is that the number of independent observations is the real cause of filter degeneracy because they make the likelihood peak in only a very small portion of the observation space. However, no matter what the cause is, more is needed than simple resampling to solve the degeneracy problem.

4.2   The Auxiliary Particle Filter

Up to now we only considered resampling at the time of the actual observation, and one of the conclusions is that this is already too late: the weights of the particles are too diverse to avoid filter degeneracy. Indeed, as soon as we arrive at observation time and the weights are degenerate the particle filter cannot be saved: the information to save it is just not there. The only thing one can do is start the runs over again, perhaps in a smarter way.

So the question is, is it possible to do something to the particles before observation times? The answer is yes, and several schemes have been proposed. Perhaps the first idea that comes to mind is to run the particles towards the new observations, see how they are doing, and redo the runs with better positioned particles. This is indeed what the Auxiliary Particle Filter does, see Pitt and Shephard (1999) for details. The scheme runs as follows (see Figure 4):

1 Integrate each particle from the previous observations at time $n - m$ to the new observations at time n. This can be done with simplified dynamics (e.g. without model noise) as the reason for these runs is to probe where the observations are.
2 Weight each particle with the new observations as

$$\tilde{w}_i \propto p(y^n | x_i^n) \tag{1.49}$$

where we assumed that the weights are equal at time $n - m$. These weights are called the 'first-stage weights' or the 'simulation weights'.
3 Resample the particles at time $n - m$ with these weights, and use these resampled particles k_i as second part of the proposal density by integrating each forward to n with the full stochastic model, so choosing from $p(x_{k_i}^j | x_{k_i}^{j-1})$. Note that k_i connects the original particle i with its new position in state space, that of particle k from the resampling step at $n - m$.
4 Re-weight the members with weights

$$w_i^n = \frac{1}{A} \frac{p(y^n | x_i^n)}{\tilde{w}_{k_i}} \tag{1.50}$$

in which A is the normalisation factor. A resampling step can be done, but is not really necessary because the resampling is done at step 3.

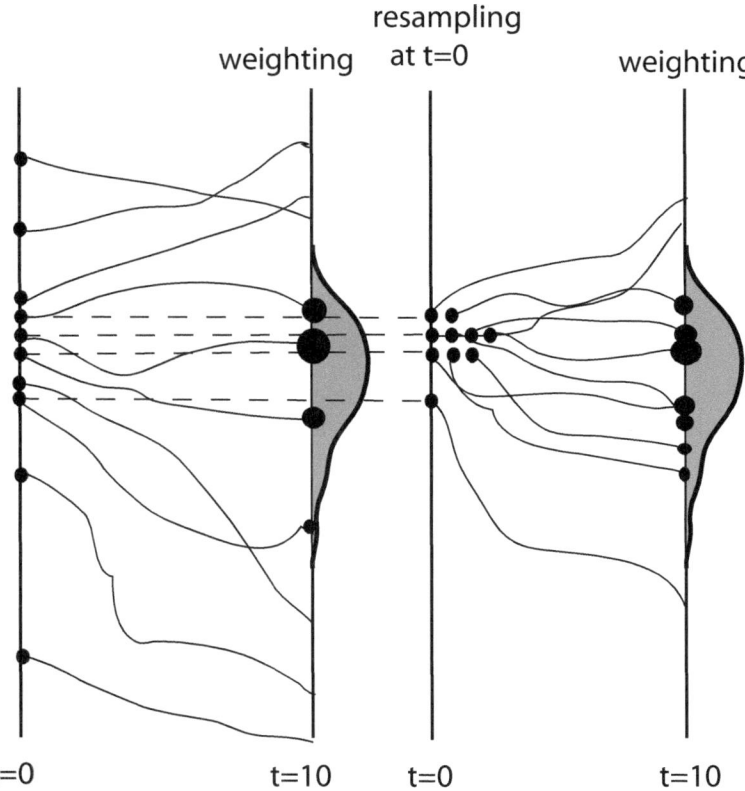

Fig. 4 *The Auxiliary Particle Filter. The model variable runs along the vertical axis, the weight of each particle corresponds to the size of the bullets on this axis. The horizontal axis denotes time, with observations at a time interval of 10 time units. All particles have equal weight at time zero. At time 10 the particles are weighted according to the likelihood. These weights are used at time 0 to rerun the ensemble up to time 10.*

The name 'auxiliary' comes from the introduction of the member index k_i in the formulation. This member index keeps track of the relation between the first-stage weights and the particle sample at observation time $n - m$.

A drawback of this scheme is that 2N integrations have to be performed, one ensemble integration to find the proposal, and one for the actual pdf. If adding the stochastic noise is not expensive, step 1 can be done with the stochastic model, which comes down to doing Sequential Importance Resampling twice. However, one could also use a simplified model for the first set of integrations. A geophysical example would be to use a simplified model like a quasi-geostrophic model for the first set of integrations, and the full model for the second.

It is possible to do it even more times, zooming in into the likelihood, but at a cost of performing more and more integrations of the model. Unfortunately, my experience with this method is not encouraging for high-dimensional geophysical applications. Other methods based on similar principles, like the Guiding Particle Filter which employs resampling before the actual observations with increased observational errors, have been proposed, see the review by Van Leeuwen (2009), but again experience shows that these methods do not perform well on high-dimensional systems. The conclusion is that one has to 'tell the particles where to go', so give them information on where future observations are. This is indeed possible, as explored in the next section.

4.3   Localisation in particle filters

As we have seen (and will become even more clear later on) the problem with particle filters is the number of observations that form the likelihood. If we could somehow reduce this number particle filters would perform much better. One option is to combine different observations, sometimes called superobbing or summary statistics, but that will reduce the information in the observations, while typically the system is under observed from the start.

Another option is to split the observations into different local in space batches, and use separately Bayes Theorem on these different areas. This process is called localisation. It is a standard technique in Ensemble Kalman Filters. As will be discussed later, Ensemble Kalman Filters approximate the Kalman Filter by representing the mean and covariance of the prior pdf with a small number of model states, typically in the order of 10–500. In the real full covariance correlations between distant points in space will be very close to zero. However, when we try to represent the covariance with let's say 100 states, spurious long-range correlations will result: it is unlikely that 100 random numbers drawn from a zero-mean density add up to zero. To avoid this problem one cuts of the covariance after some distance, either smoothly, or via a hard cut off. In this way any point will only see the observations within this so-called localisation radius. In Ensemble Kalman Filters this procedure also helps with the conditioning of the data assimilation, but that is less of an issue here.

We could do the same in particle filtering. The issue, as pointed out by Van Leeuwen (2009), is that the weight of each particle will vary over space. The issue comes when trying to resample. Particle 1 might do very good in one area, so have a high weight there, and have a very low weight in another area. Should that particle be resampled or not? One option is to only resample it when where its weight is high. But that would mean that we have to somehow stitch particles from different areas together for the following forward integration of the model. Geophysical models are generally very sensitive to artificial large gradients, typically leading to unphysical behaviour. So this direct application of localisation is not possible.

However, recent developments using optimal transportation allow for smoother updates (Reich, 2013). The idea is that at each grid point each new particle can we written as a linear combination of the old particles. By making the coefficients smooth the result will be a smooth new particle. Although the new particles will be smooth in time, that does not necessarily mean they are also 'balanced', i.e. fulfil physical relations between different model variables to a large extent, see introduction. This idea has not been explored in geophysical systems yet, but is definitely of interest and needs further exploration.

Interestingly, using proposal densities is discussed in the next sections implicitly also induces localisation as each grid point sees only observations within an influence region set by the covariance matrix used in the proposal density. So localisation is fully explored in more advanced particle filters.

5 Proposal densities

In this section we will concentrate on recent developments in using the so-called proposal transition density in solving the degeneracy problem.

First, we discuss how a proposal density is used in particle filtering, and discuss as simple example the Ensemble Kalman filter at the time of the present observations to move particles towards these observations before the likelihood weights are calculated. In the next chapter we discuss methods that change the model equations by informing intermediate steps between observations about the future observations to steer them in the right direction during the model integrations.

5.1 Proposal densities: theory

We are now to discuss a very interesting property of particle filters that is explored more and more in the geophysical community. As always we start from Bayes Theorem:

$$p(x^{0:n}|y^{0:n}) = \frac{p(y^n|x^n)p(x^n|x^{n-1})}{p(y^n)}p(x^{0:n-1}|y^{1:n-1}) \tag{1.51}$$

To simplify the analysis, and since we concentrate on a filter here, let us first integrate out the past, to get:

$$p(x^n|y^{0:n}) = \frac{p(y^n|x^n)}{p(y^n)}\int p(x^n|x^{n-1})p(x^{n-1}|y^{1:n-1})\,dx^{n-1} \tag{1.52}$$

This expression does not change when we multiply and divide the integrant by a pdf $q(x^n|x^{n-1}, y^n)$ called the proposal transition density, so:

$$p(x^n|y^{0:n}) = \frac{p(y^n|x^n)}{p(y^n)}\int \frac{p(x^n|x^{n-1})}{q(x^n|x^{n-1}, y^n)}q(x^n|x^{n-1}, y^n)p(x^{n-1}|y^{1:n-1})\,dx^{n-1}$$

$$\tag{1.53}$$

As long as the support of $q(x^n|x^{n-1}, y^n)$ is equal to or larger than that of $p(x^n|x^{n-1})$ we can always do this. This ensures we don't divide by zero. Assuming we have an equal-weight ensemble of particles from the previous analysis at time $n - 1$:

$$p(x^{n-1}|y^{1:n-1}) = \sum_{i=1}^{N} \frac{1}{N} \delta(x^{n-1} - x_i^{n-1}) \qquad (1.54)$$

we find:

$$p(x^n|y^{0:n}) = \sum_{i=1}^{N} \frac{1}{N} \frac{p(y^n|x^n)}{p(y^n)} \frac{p(x^n|x_i^{n-1})}{q(x^n|x_i^{n-1}, y^n)} q(x^n|x_i^{n-1}, y^n) \qquad (1.55)$$

As a last step, we run the particles from time $n - 1$ to n, i.e. we sample from the transition density. However, instead of drawing from $p(x^n|x_i^{n-1})$, so running the original model, we sample from $q(x^n|x_i^{n-1}, y^n)$, so from a modified model.

As an example, let us write this modified model as

$$x^n = g(x^{n-1}, y^n) + \hat{\beta}^n \qquad (1.56)$$

If we assume that $\hat{\beta}^n$ is Gaussian distributed with zero mean and covariance $\hat{Q}$, so $\hat{\beta}^n \sim N(0, \hat{Q})$ we find for the proposed transition density:

$$q(x^n|x^{n-1}, y^n) = N(g(x^{n-1}, y^n), \hat{Q}) \qquad (1.57)$$

so, as for the original transition density, the mean is the deterministic model, and the covariance is that of the stochastic term. But, of course, we don't have to use this Gaussian form, we can do 'whatever we want'.

Back to the general case, drawing from the proposal transition density leads to:

$$p(x^n|y^{0:n}) = \sum_{i=1}^{N} \frac{1}{N} \frac{p(y^n|x_i^n)}{p(y^n)} \frac{p(x_i^n|x_i^{n-1})}{q(x_i^n|x_i^{n-1}, y^n)} \delta(x^n - x_i^n) \qquad (1.58)$$

so the posterior pdf at time n can be written as:

$$p(x^n|y^{1:n}) = \sum_{i=1}^{N} w_i \delta(x^n - x_i^n) \qquad (1.59)$$

with weights w_i given by:

$$w_i \propto p(y^n|x_i^n) \frac{p(x_i^n|x_i^{n-1})}{q(x_i^n|x_i^{n-1}, y^n)} \qquad (1.60)$$

where we dropped the factors $1/N^2$ and $p(y^n)$ as they are the same for each particle. We recognise the first factor in this expression as the likelihood weight, and the second as a factor related to using the proposal transition density instead of the original transition density to propagate from time $n-1$ to n. This last factor is related to the use of the proposed model instead of the original model. These equations form the basis for exploring proposal densities to find efficient particle filters.

Finally, let us formulate an expression for the weights when multiple model time steps are present between observation times. Assume the model needs m time steps between observations.

We will explore the possibility of a proposal density at each model time steps, so for the original model we write

$$p(x^n|y^{0:n}) = \frac{p(y^n|x^n)}{p(y^n)} \int \prod_{j=n-m+1}^{n} p(x^j|x^{j-1})p(x^{n-m}|y^{1:n-m}) \, dx^{n-m:n-1} \qquad (1.61)$$

where we used the Markov property of the model. Introducing a proposal transition density at each time step we find:

$$p(x^n|y^{0:n}) = \frac{p(y^n|x^n)}{p(y^n)} \int \prod_{j=n-m+1}^{n} \frac{p(x^j|x^{j-1})}{q(x^j|x^{j-1},y^n)} q(x^j|x^{j-1},y^n)p(x^{n-m}|y^{1:n-m}) \, dx^{n-m:n-1} \qquad (1.62)$$

Assume that the ensemble at time $n - m$ is an equal weight ensemble, so

$$p(x^{n-m}|y^{1:n-m}) = \sum_{i=1}^{N} \frac{1}{N}\delta(x^{n-m} - x_i^{n-m}) \qquad (1.63)$$

and choosing randomly from the transition proposal density $q(x^j|x^{j-1},y^n)$ at each time step leads to:

$$w_i \propto p(y^n|x_i^n) \prod_{j=n-m+1}^{n} \frac{p(x_i^j|x_i^{j-1})}{q(x_i^j|x_i^{j-1},y^n)} \qquad (1.64)$$

So we find that the weights are multiplied by a p/q term each time step in which we use a modified model.

5.2 Moving particles at observation time

One can run the particles to observation time and try to improve the likelihood weights by a proposal density, as in:

$$p(x^n|y^n) = \frac{p(y^n|x^n)}{p(y)} \frac{p(x^n)}{q(x^n|y^n)} q(x^n|y^n) \qquad (1.65)$$

The issue is that one has to be able to evaluate the p/q term, which needs the pdf of both. We do have freedom on q, which we can choose to be a Gaussian for instance, but $p(x^n)$ is not at our disposal. Approximating it with e.g. a Gaussian with mean and covariance determined from the ensemble of particles would reduce the method to standard Gaussian data assimilation, which is what we want to avoid.

So, to make progress we have to take the last time step before observations into account as we do know the pdf of $p(x^n | x_i^{n-1})$. The next sections will use the ensemble Kalman filter as proposal density using this last time step in the proposal.

5.2.1 The Ensemble Kalman Filter

The Ensemble Kalman Filter (EnKF) is a well-known and often used data-assimilation method on its own. Introduced by Evensen (1994, and corrected by Burgers et al. 1998 and Houtekamer and Mitchell, 1998), it is probably the most-used data assimilation method in the geosciences because of its simplicity. The basic idea is that the prior is assumed to be Gaussian, and also the observation errors are Gaussian, with only weakly nonlinear observation operator. In that case an extremely simple algorithm arises that has influenced the field in a hard to overestimate way. As we are dealing with nonlinear filtering in this paper, the assumption is that the prior is significantly non-Gaussian, so we will not treat this method as a separate data-assimilation method here, but instead only refer to it as a possible proposal density in a particle filter.

The Kalman filter can be derived from Bayes theorem by assuming Gaussian prior and likelihood, and a linear observation operator. This then leads to:

$$
p(x|y) = \frac{p(y|x)p(x)}{\int p(y|x)p(x)\,dx}
$$

$$
\propto \exp\left[-\frac{1}{2}(x^n - \bar{x}_f^n)^T P_f^{-1}(x^n - \bar{x}_f^n) - \frac{1}{2}(y - Hx)^T R^{-1}(y - Hx) \right]
$$

$$
\propto \exp\left[-\frac{1}{2}(x^n - \bar{x}_a^n)^T P_a^{-1}(x^n - \bar{x}_a^n) \right] \tag{1.66}
$$

in which $\bar{x}_a^n$ is found from completing the squares on the variable x^n as:

$$
\bar{x}_a^n = \bar{x}_f^n + K(y - H\bar{x}_f^n) \tag{1.67}
$$

in which K is the Kalman gain given by:

$$
K = P_f^n H^T (H P_f^n H^T + R)^{-1} \tag{1.68}
$$

The posterior covariance is found as:

$$
P_a^n = (1 - KH)P_f^n \tag{1.69}
$$

This posterior mean and covariance describe the posterior covariance completely. The specialty of the Kalman filter comes from the fact that the mean and the covariance are propagated between observations with the—assumed linear—model equations:

$$x_f^n = F x_a^{n-1} \tag{1.70}$$

and

$$P_f^n = F P_a^{n-1} F^T + Q \tag{1.71}$$

in which Q is the covariance matrix of the model errors, which are assumed to be distributed according to $N(0, Q)$. These equations can be derived from the Kolmogorov or Fokker-Planck equation as the evolution equations for mean and covariance under linear dynamics (see, e.g., Jazwinski, 1970). When the model equations are nonlinear the Kalman Filter is still used as the Extended Kalman Filter in which the model is linearised. However, it can be shown that the evolution equation for the error covariance can be unstable in this approximation. Furthermore, when the dimension of the model state is very large, say 10^9, the Kalman filter covariance becomes too large to be stored.

These two considerations led to the development of the Ensemble Kalman Filter (EnKF). The Ensemble Kalman Filter (EnKF) was introduced by Evensen (1994) and modified by Burgers et al. (1998) to correct for a too narrow posterior ensemble, see also Houtekamer and Mitchell (1988). The original formulation by Burgers et al. (1998) considered the EnKF an ensemble approximation to the full pdfs involved. Between observations the pdf evolution is approximated by the evolution of its ensemble representation, in which each member evolves according to the stochastic model equations:

$$x_i^n = f(x_i^{n-1}) + \beta_i^n \tag{1.72}$$

Since each member uses the full nonlinear model the instability problem in the Extended Kalman filter is eliminated. At observation time each ensemble member is forced to follow the transformation of the mean in the Kalman filter, with a few small modifications:

$$x_{i,a}^n = x_{i,f}^n + K^e \left(y - H(x_{i,f}^n) - \epsilon_i \right) \tag{1.73}$$

(This update scheme is called the stochastic EnKF in recent literature.) The first modification is that the model equivalent of the observations, $H(x_{i,f}^n)$, is perturbed by random vector drawn from the observation error pdf $N(0, R)$. The idea behind this is that each ensemble member is statistically indistinguishable from the true evolution of the system. Since the observation is a direct measurement of the true state perturbed by the observation error, each ensemble member is treated in the same way. (Note that Burgers et al. (1998) use a more pragmatic argument.) And indeed, in using this extra perturbation the posterior ensemble has a covariance which is statistically identical to that to the full Kalman filter. The second modification is the use of the full nonlinear observation operator H in the equation above.

More generally, when the observation operator is nonlinear three approaches can be followed. In the first approach the observation operator is linearised around the present forecast. More sophisticated is the second approach in which the state vector is augmented with the model equivalent of the observations with a nonlinear observation operator, so the new state vector becomes $x_i^{new} = [x^T, H(x)^T]^T$. Observing this state vector leads to a linear observatory operator, and we can use the formalism outlined above. (Note that the relations between the model state x and the observed model state $H(x)$ is nonlinear, so the prior will not be Gaussian, but that assumption is still made, consistent with the Kalman filter idea. Note also that this is allowed here as it is just a way to generate proposal samples, not to find a best solution to the data-assimilation problem.)

The third approach is to find the mode of the posterior for each ensemble member assuming that each member has a Gaussian prior with covariance determined directly (or after localisation and inflation) from the ensemble (Maximum Likelihood Ensemble Filter by Zupanski, 2005, who used a square-root formulation, see also the Ensemble Randomised Maximum Likelihood Filter of Gu and Oliver (2007), who use the stochastic EnKF). This mode can be found as

$$x_{i,a}^n = argmin \left(\frac{1}{2}(x - x_{i,f}^n)^T P_f^{-1}(x - x_{i,f}^n) + \frac{1}{2}(y^n - H(x)^T R^{-1}(y^n - H(x)) \right)$$
(1.74)

The solution to this minimisation problem is found iteratively. This method is called the Maximum Likelihood Ensemble Filter (MLEF), while the method searches for the maximum a posteriori state. It has been shown to outperform the EnKF when the observation operator is strongly nonlinear. A second modification is that the observation operator can be nonlinear, and is linearised only in the ensemble Kalman filter gain Matrix $K^e = P_f^e H^T (H P_f^e H^T + R)^{-1}$, with the superscript e denoting the ensemble representation of the covariance.

Before we discuss the EnKF as proposal density four issues should be noted. Firstly, the perturbation of the innovation by ϵ_i leads to extra sample variance, which is unwanted. Several variants of the EnKF have been derived that avoid these extra perturbations (see, e.g., Tippett et al., 2003), all based on square-root approximations of the full Kalman filter.

Secondly, the algorithm depicted above is not very efficient, for instance it looks as if the full error covariance needs still to be stored. Several efficient methods have been developed to avoid this storage problem. They are all based on ensemble representations of the covariance matrix. Define the ensemble perturbation matrix from the ensemble mean as:

$$X_f = \frac{1}{\sqrt{N-1}}(x_{1,f} - \bar{x}_f, \ldots, x_{N,f} - \bar{x}_f)$$
(1.75)

in which the first subscript denotes the ensemble member number. The prior covariance can now be written as

$$P_f = X_f X_f^T \tag{1.76}$$

The calculations needed in the Kalman gain matrix are

$$P_f H^T = X_f X_f^T H^T = X_f \left(H X_f \right)^T \tag{1.77}$$

showing that we need only to store matrices of order *model state dimension times observation dimension*. Furthermore

$$H P_f H^T = H X_f X_f^T H^T = \left(H X_f \right) \left(H X_f \right)^T \tag{1.78}$$

is only of size *observation dimension times observation dimension*.

Thirdly, if the ensemble size is N, the rank of the covariance matrix is at most $N - 1$, so all observations can only express themselves in an $N - 1$ dimensional space, which is quite restrictive if the number of observations is large. Furthermore, model covariances tend to drop to zero when the distance between two points is large. However, when the covariance is constructed using only a limited ensemble size it is unlikely that zero covariances are indeed represented as zero. The issue is that one needs a large number of random numbers drawn from a pdf with mean zero to get a zero mean. This has led to a technique called *localisation* in which covariances over large distances are set to zero. Several localisation methods exist, but no best method has been found. One of the issues is that geophysical systems often exhibit subtle relations between model variables (so-called *model balances*) that can easily be destroyed by localisation.

Finally, the finite ensemble size means that only a finite space is spanned by the ensemble members. This means that the covariance matrix will always be an underestimate of the true covariance, potentially leading to filter divergence in which the ensemble mean deviates strongly from the truth but the ensemble error covariance remains small. A very crude solution is to increase all covariances in the covariance matrix by a fixed number called the *inflation factor*. The crudeness is related to the fact that the covariance is inflated in the directions in which we do have covariance information from the ensemble, not in the missing directions. Nevertheless, this procedure does lead to much better performance of the EnKF. See, e.g., the book by Evensen (2009) for more details on these matters.

5.2.2 The Ensemble Kalman Filter as proposal density

As a first example we will explore proposal densities with the Gaussian of the EnKF as the proposal density, as proposed by Papadakis et al. (2010). They call it the Weighted Ensemble Kalman Filter (WEnKF). We know from the chapter on resampling that waiting until the observation time is too late. So we have to combine the last time step of the model before observation time with the EnKF algorithm. Let us assume we have used the EnKF at observation time, so we know what has happened in this last time step: the model is propagated forward with the model equations and the particles are moved around during the EnKF analysis step. So we know the starting point of the simulation, x_i^{n-1}, and its end point, the posterior EnKF sample x_i^n, and we know the model equation, written formally as:

$$x_i^n = f(x_i^{n-1}) + \beta_i^n \tag{1.79}$$

Hence we can determine β_i^n from this equation directly. We also know the distribution from which this β_i^n is supposed to be drawn, let us say a Gaussian with zero mean and covariance Q. We then find for the transition density:

$$p(x_i^n|x_i^{n-1}) \propto \exp\left[-1/2 \left(x_i^n - f(x_i^{n-1})\right)^T Q^{-1} \left(x_i^n - f(x_i^{n-1})\right)\right] \tag{1.80}$$

This will give us a number for each $[x_i^{n-1}, x_i^n]$ combination, so each particle. (Note that Papadakis et al., 2010, and also Beyou et al., 2013 make the mistake equating this with the proposal transition density, so they ignore the p/q term from the proposal. This action leads to less variance in the weights, so better performance of the scheme. Livings, private communication)

Let us now calculate the value of the proposal density $q(x_i^n|x_i^{n-1}, y^n)$. This depends on the ensemble Kalman filter used. For the stochastic Ensemble Kalman filter the situation is as follows. Each particle in the updated ensemble is connected to its forecast analysis as:

$$x_i^n = x_i^{n,old} + K^e \left(y - \epsilon_i - H(x_i^{n,old})\right) \tag{1.81}$$

in which ϵ_i is the random error drawn from $N(0, R)$ as explained before. The particle prior to the analysis comes from that of the previous time step through the stochastic model:

$$x_{i,f}^n = f(x_i^{n-1}) + \beta_i^n \tag{1.82}$$

Combining these two gives:

$$x_{i,a}^n = f(x_i^{n-1}) + \beta_i^n + K^e \left(y - \epsilon_i - H(x_i^{n-1}) - H(\beta_i^n))\right) \tag{1.83}$$

or

$$x_{i,a}^n = f(x_i^{n-1}) + K^e \left(y - H(f(x_i^{n-1}))\right) + (1 - K^e H)\beta_i^n - K^e \epsilon_i \tag{1.84}$$

assuming that H is a linear operator. The right-hand side of this equation has a deterministic and a stochastic part. The stochastic part provides the spread in the transition density going from x_i^{n-1} to x_i^n. Assuming both model and observation errors to be independent Gaussian distributed we find for this transition density:

$$q(x_i^n | x_i^{n-1} y^n) \propto \exp\left[-1/2 \left(x_i^n - \mu_i^n\right)^T \Sigma_i^{-1} \left(x_i^n - \mu_i^n\right)\right] \tag{1.85}$$

in which μ_i^n is the deterministic 'evolution' of x, given by:

$$\mu_i^n = f(x_i^{n-1}) + K^e \left(y - H(x_i^{n-1})\right) \tag{1.86}$$

and the covariance Σ_i is given by:

$$\Sigma_i = (1 - K^e H)Q(1 - K^e H)^T + K^e R K^{eT} \tag{1.87}$$

where we assumed as usual that the model and observation errors are uncorrelated. It should be realised that x_i^n does depend on all $x_{j,f}^n$ via the Kalman gain, that involves the error covariance P^e. Hence we have actually calculated $q(x_i^n | P^e, x_i^{n-1}, y^n)$ instead of $q(x_i^n | x_i^{n-1}, y^n)$, in which P^e depends on all other particles. The reason why we ignore the dependence on P^e is that in case of a very large ensemble P^e would be a variable that depends only on the system, not on specific realisations of that system. This is different from the terms related to x_i^n, that will depend on the specific realisation for β_i^n even when the ensemble size is very large. (Hence another approximation related to the finite size of the ensemble comes into play here and at this moment it is unclear how large this approximation error is.)

The calculations of $p(x^n | x^{n-1})$ and $q(x_i^n | x_i^{n-1} y^n)$ look like very expensive operations. By realising that Q and R can be obtained from the ensemble of particles, computationally efficient schemes can easily be derived.

We can now determine the full new weights, again ignoring normalisation factors as they are the same for each particle. The full procedure is as follows (see Figure 5):

1 Run the ensemble up to the observation time
2 Perform a (local) EnKF analysis of the particles
3 Calculate the proposal weights $w_i^* = p(x_i^n | x_i^{n-1}) / q(x_i^n | x_i^{n-1} y^n)$
4 Calculate the likelihood weights $w_i = p(y^n | x_i^n)$
5 Calculate the full relative weights as $w_i = w_i * w_i^*$ and normalise them.
6 Resample

It is good to realise that the EnKF step is only used to draw the particles close to the observations. Hence if the weights are still varying too much, one can do the EnKF step with much smaller observational errors, or do it several times. This might look like over fitting but it is not since the only thing we do in probabilistic sense is to move particles to those positions in state space where the likelihood is large and all tricks we do with the EnKF are just part of the proposal, and can be compensated for by using the correct weights.

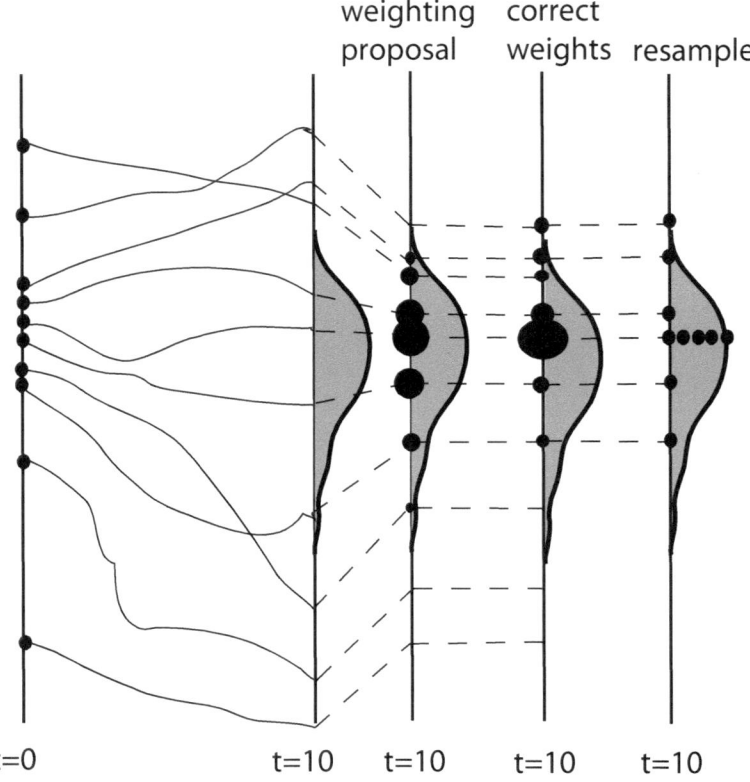

weighting   correct
proposal   weights resample

t=0   t=10   t=10   t=10   t=10

Fig. 5 *The particle filter with proposal density. The model variable runs along the vertical axis, the weight of each particle corresponds to the size of the bullets on this axis. The horizontal axis denotes time, with observations at a time interval of 10 time units. All particles have equal weight at time zero. At time 10 the particles are brought closer to the observations by using, e.g., the EnKF. Then they are weighted with the likelihood and these weights are corrected for the artificial EnKF step.*

Unfortunately, as mentioned above, there is a problem with the WEnKF as proposed by Papadakis et al. (2010). When this mistake is corrected for it becomes clear that this method does not work in high-dimensional systems (Livings, private communication). Again, more is needed, and the next chapter shows ways to move forward.

6 Changing the model equations

In this chapter we will explore particle filters in which the model equations are changed directly, fully exploring the proposal density freedom. We have in fact done this earlier with the EnKF as proposal, but there we used the EnKF equations to define the proposal density in the last time step before the observations. And in the auxiliary particle filter one could use a different model for the first set of integrations to obtain the first-stage weights, but the future observations were not used directly in the model equations. Much more efficient schemes can be derived that change the model equations such that each particle is pulled towards the future observations at each time step, and that is what we will explore in this chapter. By keeping track of the weights associated with this it can be assured that the correct problem is solved, and the particles remain random samples from the posterior pdf.

As mentioned before, the idea of the proposal transition density is that we draw samples from that density instead of from the transition density related to the original model. And the efficiency gain exists in the fact that these samples can be dependent on the future observations. To see how this works consider the following example. Write the model equations as:

$$x_i^j = f(x_i^{j-1}) + \beta_i^j \tag{1.88}$$

Let us recall how this equation is related to the transition density of the original model, $p(x_i^j|x_i^{j-1})$. The probability to end up in state x_i^j starting from particle position x_i^{j-1} is related to β_i^j. For instance, if $\beta_i^j = 0$, so no model error, or a so-called perfect model, this probability is 1 if the x_i^j, x_i^{j-1} pair fulfils the perfect model equations $x_i^j = f(x_i^{j-1})$, and zero otherwise. In this case $p(x_i^j|x_i^{j-1})$ is be a delta function centred on $f(x_i^{j-1})$. In the more realistic case the model error is nonzero and the transition density will depend on the distribution of the stochastic random forcing. Assuming Gaussian random forcing with mean zero and covariance Q, so $\beta_i^j \sim N(0, Q)$, we find

$$p(x_i^j|x_i^{j-1}) \propto N(f(x_i^{j-1}), Q) \tag{1.89}$$

so the mean of the Gaussian is the deterministic move, and the covariance of the transition density is determined by the covariance of the stochastic forcing.

Let us now explore a modified model equation, one that 'knows' about future observations, and actually draws the model to those observations. A very simple example is to add a term that relaxes the model to the future observation, like

$$x_i^j = f(x_i^{j-1}) + \beta_i^j + T^j(y^n - H(x_i^{j-1})) \tag{1.90}$$

in which n is the next observation time. The observation operator H does not contain any model integrations, it is just the evaluation of x_i^{j-1} in observation space, simply because x_i^n is not known yet. Clearly, each particle i will now be pulled towards the future observations, with relaxation strength related to matrix T^j. In principle, we are free to choose this matrix, but it is reasonable to assume that it is related to the error covariance of the future observation R, and of the model equations Q. Possible forms will be discussed in the examples later.

Of course, we can't just change the model equations, we have to compensate for this change via the weight of the particles, as explained in chapter 4, where we showed that the proposal density turns up in the weights. Each time step the weight of each particle changes with

$$w_i^j = \frac{p(x_i^j | x_i^{j-1})}{q(x_i^j | x_i^{j-1}, y^j)} \tag{1.91}$$

between observation times. This can be calculated in the following way. Using the modified model equations, we know the particle position x_i^{j-1} for each particle, that was our starting point, and also x_i^j. So, assuming the model errors are Gaussian distributed, this would make

$$p(x_i^j | x_i^{j-1}) \propto \exp\left[-\frac{1}{2} \left(x_i^j - f(x_i^{j-1}) \right)^T Q^{-1} \left(x_i^j - f(x_i^{j-1}) \right) \right] \tag{1.92}$$

The proportionality constant is not of interest since it is the same for each particle, so it drops out when the relative weights of the particles are calculated. Hence we have all ingredients to calculate this, and $p(x_i^j | x_i^{j-1})$ is just a number.

For the proposal transition density we use the same argument, to find:

$$
\begin{aligned}
q(x_i^j | x_i^{j-1} y^n) &\propto \exp\left[-\frac{1}{2} \left(x_i^j - f(x_i^{j-1}) - T^j(y^n - H(x^{j-1})) \right)^T Q^{-1} \left(x_i^j - f(x_i^{j-1}) - T^j(y^n - H(x^{j-1})) \right) \right] \\
&= \exp\left[-\frac{1}{2} \beta_i^{jT} Q^{-1} \beta_i^j \right]
\end{aligned}
\tag{1.93}
$$

Again, since we have already chosen β to propagate the model state forward in time, we can calculate this and it is just a number. In this way, any modified equation can be used, and we know, at least in principle, how to calculate the appropriate weights.

6.1 The 'Optimal' proposal density

In the literature the so-called optimal proposal density is defined (see, e.g., Doucet et al., 2001) by taking $q(x^n | x^{n-1}, y^n) = p(x^n | x^{n-1}, y^n)$. It is argued this choice results in optimal weights, and explains the word 'optimal'. However, it is easy to show that particle filters would never be useful for high-dimensional systems if this choice led to optimal weights, as shown below. Let us study the simplest case. Assume observations every time step, and a resampling scheme at every time step, so that a equal-weighted ensemble of particles is present at time $n-1$. Furthermore, assume that model errors are Gaussian distributed as $N(0, Q)$ and observation errors are Gaussian distributed according to $N(0, R)$. First, using Bayes Theorem we can write:

$$p(x^n | x^{n-1}, y^n) = \frac{p(y^n | x^n) p(x^n | x^{n-1})}{p(y^n | x^{n-1})} \tag{1.94}$$

where we used $p(y^n|x^n, x^{n-1}) = p(y^n|x^n)$, i.e. we don't learn anything new for the pdf of the observations from x^{n-1} when we know x^n. Using this proposal density gives posterior weights:

$$w_i = p(y^n|x_i^n)\frac{p(x_i^n|x_i^{n-1})}{q(x_i^n|x_i^{n-1}, y^n)}$$

$$= p(y^n|x_i^n)\frac{p(x_i^n|x_i^{n-1})}{p(x_i^n|x_i^{n-1}, y^n)}$$

$$= p(y^n|x_i^{n-1}) \tag{1.95}$$

To evaluate this term we can expand it as:

$$w_i = \int p(y^n, x^n|x^{n-1}) \, dx^n = \int p(y^n|x^n)p(x^n|x^{n-1}) \, dx^n \tag{1.96}$$

in which we again used $p(y^n|x^n, x^{n-1}) = p(y^n|x^n)$. Using the Gaussian assumptions mentioned above (note that the state is never assumed to be Gaussian), we can perform the integration to obtain:

$$w_i \propto \exp\left[-\frac{1}{2}\left(y^n - Hf(x_i^{n-1})\right)^T (HQH^T + R)^{-1}\left(y^n - Hf(x_i^{n-1})\right)\right] \tag{1.97}$$

We will now obtain an order of magnitude estimate for the variance of $-\log w_i$ as function of i. First expand $y - Hf(x_i^{n-1})$ to $y - Hx_t^n + H\left(x_t^n - f(x_i^{n-1})\right)$ in which x_t^n the true state at time n. If we now use $x_t^n = f(x_t^{n-1}) + \beta_t^n$, this can be expanded further as $y - Hx_t^n + H\left(f(x_t^{n-1}) - f(x_i^{n-1})\right) + H\beta_t^n$. To proceed we make the following restrictive assumptions that will nevertheless allow us to obtain useful order-of-magnitude estimates. Let us assume that both the observation errors R and the observed model errors HQH^T are uncorrelated, with variances V_y and V_β, respectively, to find:

$$-\log(w_i) = \frac{1}{2(V_\beta + V_y)}\sum_{j=1}^{M}\left[y_j - H_jx_t^n + H_j\beta_t^n + H_j\left(f(x_t^{n-1}) - f(x_i^{n-1})\right)\right]^2 \tag{1.98}$$

The variance of $-\log w_i$ arises from varying ensemble index i. Introduce the constant $\gamma_j = y_j^n - H_jx_t^n + H_j\beta_t^n$ and assume that the model can be linearised as $f(x_i^{n-1}) \approx Ax_i^{n-1}$, leading to:

$$-\log(w_i) = \frac{1}{2(V_\beta + V_y)}\sum_{j=1}^{M}\left[\gamma_j + H_jA(x_t^{n-1} - x_i^{n-1})\right]^2 \tag{1.99}$$

A following step in our order of magnitude estimate is to assume $x_t^{n-1} - x_i^{n-1}$ to be Gaussian distributed. In that case the expression above is non-central χ_M^2 distributed apart from a constant. This constant comes from the variance of $\gamma_j + H_jA(x_t^{n-1} - x_i^{n-1})$, which is equal to $H_jAP^{n-1}A^TH_j^T$, in which P^{n-1} is the covariance of the model state at time $n-1$. Defining $V_x = H_jAP^{n-1}A^TH_j^T$, we find:

$$-\log(w_i) = \frac{V_x}{2(V_\beta + V_y)} \sum_{j=1}^{M} \frac{\left[\gamma_j + H_jA(x_t^{n-1} - x_i^{n-1})\right]^2}{V_x} \tag{1.100}$$

Apart from the constant in front the expression above is non-central χ_M^2 distributed with variance $a^2 2(M + 2\lambda)$ where $a = V_x/(2(V_\beta + V_y))$ and $\lambda = (\sum_j \gamma_j^2)/V_x$.

We can estimate an order of magnitude for λ by realising that for a large enough number of observations we expect $\sum_j (y_j^n - H_jx_t^n)^2 \approx MV_y$, and $\sum_j (y_j^n - H_jx_t^n) \approx 0$. Furthermore, when the dimension of the system under study is large we expect $\sum_j (H_j\beta_t^n)^2 \approx MV_\beta$. Combining all these estimates we find that the variance of $-\log(w_i)$ can be estimated as

$$\frac{M}{2} \left(\frac{V_x}{V_\beta + V_y}\right)^2 \left(1 + 2\left(\frac{V_\beta + V_y}{V_x}\right)\right) \tag{1.101}$$

This expression shows that the only way to keep the variance of $-\log(w_i)$ low when the number of independent observations M is large is to have a very small variance in the ensemble: $V_x \approx (V_\beta + V_y)/M$, see Figure 6. Clearly, when the number of observations is large (100 million in typical meteorological applications),

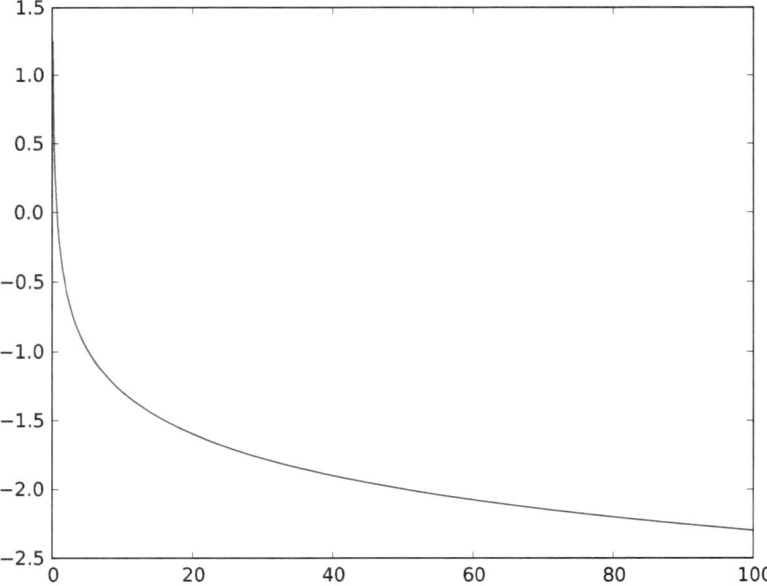

Fig. 6 $\log_{10}$ *of the factor multiplying M in the variance of $-\log w_i$ as function of $(V_\beta + V_y)/V_x$. Note that even when the latter is 100 a system with more than 1000 observations will be degenerate.*

this is not very realistic. This expression has been tested in several applications and holds within a factor 0.1 in all tests (Van Leeuwen, 2014, unpublished manuscript to be submitted).

A large variance of $-\log(w_i)$ does not necessarily mean that the weights will be degenerate because the large variance could be due to a few outliers on the low end. However, we have shown that $-\log(w_i)$ is approximately non-central χ^2_M distributed for a linear model, so the large variance is not due to outliers but intrinsic in the sampling from such a distribution. There is no reason to expect that this variance will behave better for nonlinear models, especially because we didn't make any assumptions on the divergent or contracting characteristics of the linear model.

From this analysis we learn two lessons: the number of independent observations determines the degeneracy of the filter, and the optimal proposal density cannot be used in systems with a very large number of independent observations.

6.2   The Implicit Particle Filter

The Implicit Particle Filter was introduced in Chorin and Tu (2009) and developed further in Chorin et al. (2010) and Morzfeld et al. (2012). As we shall see, the method is closely related to the 'optimal proposal density' discussed above when the observations are available at every time step.

The method works as follows. Assume we have m model time steps between observations. Define the function $F(..)$ as:

$$F_i(x^{n-m+1:n}) = -\log(p(y^n|x^n)p(x^{n-m+1:n}|x_i^{n-m})) \qquad (1.102)$$

and its minimum

$$\phi_i = \min(F_i(x^{n-m+1:n})) \qquad (1.103)$$

Draw a random vector ξ_i of length the size of the state vector times m, with each element of ξ_i is drawn from $N(0,1)$. The actual samples are now constructed by solving

$$F_i(x^{n-m+1:n}) = \frac{\xi_i^T \xi_i}{2} + \phi_i \qquad (1.104)$$

for each particle x_i. The term ϕ_i is included to ensure that the equation above has a solution. One can view this step as drawing from the proposal density $q_x(x^{n-m+1:n}|x_i^{n-m}, y^n)$ via the proposal density $q_\xi(\xi)$, where we introduced the subscript to clarify the shape of the pdf. These two are related by a transformation of the probability densities as

$$q_x(x^{n-m+1:n}|x_i^{n-m}, y^n)dx^{n-m:n} = q_\xi(\xi)d\xi \qquad (1.105)$$

so that

$$q_x(x^{n-m:n}|x_i^{n-m}, y^n) = q_\xi(\xi)J_i^{-1} \tag{1.106}$$

in which J_i is the Jacobian of the transformation $x \to \xi$. We can now write the weights of this scheme as:

$$
\begin{aligned}
w_i &= p(y^n|x_i^n)\frac{p(x_i^{n-m+1:n}|x_i^{n-m})}{q_x(x_i^{n-m+1:n}|x_i^{n-m}, y^n)} \\
&= p(y^n|x_i^n)\frac{p(x^{n-m+1:n}|x_i^{n-m})}{q_\xi(\xi_i)}J_i \\
&= \frac{\exp\left(-F_i(x^{n-m+1:n})\right)}{\exp\left(-1/2\xi^T\xi\right)}J_i \\
&= \frac{\exp(-F_i\left(x^{n-m+1:n}\right))}{\exp\left(-F_i(x^{n-m+1:n}) + \phi_i\right)}J_i = \exp(-\phi_i)\,J_i \tag{1.107}
\end{aligned}
$$

where we used (1.104).

To understand better what this means in terms of the variance of the weights let's consider the case of observations every model time step, and Gaussian observation errors, Gaussian model equation errors, and linear observation operator H. In that case we have

$$
\begin{aligned}
-\log(p(y^n|x^n)p(x^n|x_i^{n-1})) &= \frac{1}{2}\left(y^n - Hx_i^n\right)^T R^{-1}\left(y^n - Hx_i^n\right) \\
&+ \frac{1}{2}\left(x^n - f(x_i^{n-1})\right)^T Q^{-1}\left(y^n - f(x_i^{n-1})\right) \\
&= \frac{1}{2}\left(x^n - \hat{x}_i^n\right)^T P^{-1}\left(x^n - \hat{x}_i^n\right) + \phi_i \tag{1.108}
\end{aligned}
$$

in which $\hat{x}_i^n = f(x_i^{n-1}) + K(y^n - Hf(x_i^{n-1}))$, the maximum of the posterior pdf, and $P = (1 - KH)Q$, with $K = QH^T(HQH^T + R)^{-1}$. Comparing this with (1.104) we find $x_i^n = \hat{x}_i^n + P^{1/2}\xi_i$, so J is a constant, and

$$
\begin{aligned}
\phi_i &= \min(-\log(p(y^n|x^n)p(x^{n-m+1:n}|x_i^{n-m}))) \\
&= \frac{1}{2}\left(y^n - Hf(x_i^{n-1})\right)^T (HQH^T + R)^{-1}\left(y^n - Hf(x_i^{n-1})\right) \tag{1.109}
\end{aligned}
$$

and finally $w_i \propto \exp(-\phi_i)$. Comparing with the optimal proposal density we see that when observations are present at every time step the implicit particle filter is equal to the optimal proposal density, with the same degeneracy problem. Although the Implicit Particle Filter can be used with nonlinear H and more general error pdfs there is no reason to assume that it will not suffer from filter degeneracy in these more general cases too.

We have not discussed yet how to actually solve the equation

$$F_i(x^{n-m+1:n}) = \frac{\xi_i^T \xi_i}{2} + \phi_i$$

for each particle. Clearly, since F_i is a scalar this equation has an infinite number of solutions. When the model equation is linear and the observation operator H is linear too, this is a quadratic equation in trajectory $x_i^{n-m+1:n}$. In that case F_i can be rewritten as:

$$F_i(x^{n-m+1:n}) = (x_i^{n-m+1:n} - x_i^a)^T P^{-1} (x_i^{n-m+1:n} - x_i^a) + \phi_i$$

in which x_i^a and P follow from completing the squares in F_i, as:

$$x_i^a = M x_i^{n-m} + K(y - H^n M x_i^{n-m}) \tag{1.110}$$

with $M x_i^{n-m}$ generates the trajectory $(x_i^{n-m+1}, \ldots, x_i^n)$ using the deterministic model, $H^n M x_i^{n-m} = H^n x_i^n$, and K is the generalised Kalman gain given by:

$$K = B H^{nT} (H^n B H^{nT} + R)^{-1} \tag{1.111}$$

with B the trajectory of the covariance using the deterministic model and evolution equation:

$$B^j = M^{j-1} B^{j-1} M^{j-1} + Q^j \tag{1.112}$$

where Q is the error covariance of the model equations. The covariance P is given by:

$$P = (1 - KH^n)B \tag{1.113}$$

One possible solution is to choose

$$x_i^{n-m+1:n} = x_i^a + P^{1/2} \xi_i$$

If the model and/or H is nonlinear, we can either linearise the system and solve the above iteratively (see next section), or use a method called the random map, in which we assume (see Morzfeld et al.)

$$x_i^{n-m+1:n} = x_i^a + \lambda(\xi_i)\xi_i \tag{1.114}$$

in which $\lambda(\xi_i)$ is a scalar function of ξ_i. This leads to a highly nonlinear but scalar equation for λ that is typically easier to solve than solving the high-dimensional problem iteratively.

Finding ϕ_i, the minimum of F_i is close to the solution of a variational method called 4DVar, which we will discuss in the next section, followed by using 4DVar as a proposal density.

6.3 Variational methods as proposal densities

A data-assimilation technique widely used in the geosciences, especially in numerical weather forecasting, is a variational technique called 4DVar. This method tries to find the mode of the posterior pdf joint over time, so the pdf of model trajectories over a certain time window. It is interesting that this mode is chosen, and not the mode of the marginal pdf at final time. Given the nonlinear models used in numerical weather forecasting, these two modes are not the same. The reason is one of computational efficiency as will be explained below.

The prior and likelihood cannot have any shape for this method to be efficient. Typically Gaussian assumptions are made for the prior, the observation errors, and, if they are included, errors in the model equations. Because the model equations are nonlinear the observation operator, which takes the model state at time zero to the observation time and projects it to observation space, is also nonlinear. On top of that, the actual observation operator at observation time can also be a nonlinear operator. In what follows we will discuss how 4DVar is used in present-day numerical weather forecasting, followed by how 4DVar can be made a useful tool for nonlinear non-Gaussian data assimilation.

6.3.1 4DVar as stand-alone method

Let us first look at the case when model errors are ignored, and the unknown is the model state at the start of a time window, so-called *strong-constraint 4DVar*. Hence the solution to the data-assimilation problem is taken as the mode of the marginal pdf at time zero given a set of observations within the window.

Let us have a look at the algorithm in more detail, following Fisher and Auvinen (2012). The posterior can then be written as:

$$p(x^0|y) \propto p(y|x^0)p(x^0) \propto \exp(-J(x^0)) \tag{1.115}$$

in which $J(x^0)$ is given by:

$$J\left(x^0\right) = \frac{1}{2}\left(x^0 - x_b\right)^T B^{-1}\left(x^0 - x_b\right)$$

$$+ \sum_{j=1}^{M} \frac{1}{2}\left(y^j - \tilde{H}^j\left(x^0\right)\right)^T R^{-1}\left(y^j - \tilde{H}^j(x^0)\right) \tag{1.116}$$

in which $\tilde{H}^j(x^0)$ takes the model state at time zero, propagates it using the deterministic nonlinear model to the time of observation set j, and projects the resulting model state into observation space. The minimisation of $J\left(x^0\right)$ is constrained to the state following the deterministic model equations:

$$x^i = f(x^{i-1}) \tag{1.117}$$

Because the model equations are nonlinear the posterior is not Gaussian in x^0, and the cost function $J(x^0)$ is not quadratic in x^0. This means that the solution cannot be written down in closed form and numerical procedures are needed. In 4DVar, as the name suggests, the minimum of the costfunction (so the maximum of the posterior), is found through variational methods.

Instead of solving the constrained optimisation problem one usually solves the unconstrained problem using Lagrange multipliers:

$$
\begin{aligned}
J\left(x^0\right) = {} & \frac{1}{2}\left(x^0 - x_b\right)^T B^{-1}\left(x^0 - x_b\right) \\
& + \sum_{j=1}^{M} \frac{1}{2}\left(y^j - H^j\left(x^j\right)\right)^T R^{-1}\left(y^j - H^j(x^j)\right) \\
& + \sum_{i=1}^{n}\left(\lambda^i\right)^T (x^i - f(x^{i-1}))
\end{aligned}
\tag{1.118}
$$

with λ^i the Lagrange multiplier fields at times i, and $H^j(x^j)$ maps model state x^j into observation space. Note that the minimisation is now not just over x^0 but over the whole trajectory. The solution of the minimisation problems follows from the calculus of variations as:

$$\frac{\partial J}{\partial x^0} = B^{-1}\left(x^0 - x_b\right) - F^{0T}\lambda^1 = 0 \tag{1.119}$$

$$\frac{\partial J}{\partial \lambda^i} = x^i - f(x^{i-1}) = 0 \tag{1.120}$$

$$\frac{\partial J}{\partial x^i} = \lambda^i - F^{iT}\lambda^{i+1} - H^{jT}R^{-1}\left(y^j - H^j(x^j)\right) = 0 \tag{1.121}$$

$$\frac{\partial J}{\partial x^n} = \lambda^n - H^{nT}R^{-1}\left(y^n - H^n(x^n)\right) = 0 \tag{1.122}$$

in which F^i is the linearisation of $f(..)$ at time i.

This constitutes a two-point boundary value problem which is solved iteratively. One starts from a first guess for x^0 and integrates the deterministic model forward to time n. Then the adjoint equations for the adjoint variables λ are solved backwards in time from λ^n to λ^1. Finally the expression for λ^1 is used to find a new guess for x^0 and the whole process is repeated until convergence.

Common practice in numerical weather prediction is not to use minimisation methods for non-quadratic cost functions, but instead solve a series of quadratic minimisation methods using Gauss-Newton iteration. This procedure is, perhaps confusingly, called *incremental 4DVar*. Each forward-backward integration is called an *inner loop*, and an *outer loop* is an integration of the full nonlinear system. As an example, the UK Met Office runs a 4Dvar with 50 inner loops and one outer loop, and the European Centre for Medium-range Weather Forecasting ECMWF runs 50 inner loops and 2 outer loops in their 4Dvar.

A popular minimisation method in meteorology is conjugent gradient. The rate of convergence depends on the condition number of the Hessian A, which is defined as the curvature of the cost function, and is given by:

$$A = \frac{\partial^2 J}{\partial x^2} = B^{-1} + \sum_{i=1}^{n} F^{i^T} \tilde{H}^{i^T} R^{-1} \tilde{H}^i F^i \tag{1.123}$$

The condition number is given as the ratio of the largest to the smallest eigenvalue of A:

$$\kappa(A) = \frac{\mu_1}{\mu_{N_x}} \tag{1.124}$$

in which N_x is the dimension of the state vector. It measures how circular symmetric the cost function is. If $\kappa(A) = 1$ all eigenvalues are equal, the costfunction has a perfectly circular symmetric shape, and a gradient descent algorithm will find the minimum in one iteration. However, a large condition number is related to a highly elliptical cost function, and a large number of iterations will be needed for convergence. Since the condition number is typically high related to very small eigenvalues a technique called preconditioning is used. In preconditioning a coordinate transformation is used to reduce the condition number. Define

$$\chi = L^{-1}(x^0 - x_b) \tag{1.125}$$

with

$$B^{-1} = LL^T \tag{1.126}$$

to find for the cost function:

$$J\left(x^0\right) = \frac{1}{2}\chi^T \chi$$
$$+ \sum_{j=1}^{M} \frac{1}{2} \left(y^j - \tilde{H}^j \left(L\chi + x_b\right)\right)^T R^{-1} \left(y^j - \tilde{H}^j (L\chi + x_b)\right) \tag{1.127}$$

with Hessian

$$\hat{A} = 1 + L^T \sum_{i=1}^{n} F^{i^T} \tilde{H}^{i^T} R^{-1} \tilde{H}^i F^i L \tag{1.128}$$

This preconditioning will make $\mu_{N_x} = 1$, strongly decreasing the condition number.

But we can do better by exploring the eigenvalues of A. Let us write the result of preconditioning on the Hessian as:

$$\hat{A} = L^T A L \tag{1.129}$$

The best preconditioning would be $L = A^{-1/2}$, which would make all eigenvalues equal to one, resulting in only one minimisation step needed. So the question becomes, can we find an approximation of the Hessian that can easily be inverted? To answer this question let us decompose the Hessian in terms of its eigenvectors as:

$$A = \sum_{i=1}^{K} \mu_i v_i v_i^T \tag{1.130}$$

Define the preconditioning matrix using the first p eigenvectors as:

$$L^{-1} = 1 + \sum_{i=1}^{p} (\sigma_k^{1/2} - 1) v_i v_i^T \tag{1.131}$$

which leads to

$$\hat{A} = L^T A L = \sum_{i=1}^{p} \sigma_i \mu_i v_i v_i^T + \sum_{i=p+1}^{N_x} \mu_i v_i v_i^T \tag{1.132}$$

Now choosing $\sigma_i \mu_i < \mu_{p+1}$ leads to a condition number

$$\kappa(\hat{A}) = \frac{\mu_{p+1}}{\mu_{N_x}} = \mu_{p+1} \tag{1.133}$$

The eigenvectors and eigenvalues can be found from the Lanczos algorithm, which is used to minimise the cost function and at the same time calculate the eigenstructure of the Hessian. This is used by, e.g., ECMWF to calculate the eigenstructure in the first inner loops and use them in the next inner loops, resulting in super linear convergence for the second inner loop.

Most centres are now moving away from strong-constrained 4Dvar and are working to include model errors, in a procedure called *weak-constrained 4Dvar*. The posterior now becomes:

$$p(x|y) \propto p(y|x)p(x) \tag{1.134}$$

in which $x = (x^0, \ldots, x^n)^T$ denotes a whole trajectory of the model. The model equations are written as:

$$x^j = f(x^{i-j}) + \beta^j \tag{1.135}$$

in which β^j is the stochastic part of the model equation at time j. Typically, but not always, β is taken white in time, leading to a Markov process and the prior pdf can be written as:

$$p(x) = p(x^n|x^{n-1})p(x^{n-1}|x^{n-2}), \ldots, p(x^1|x^0)p(x^0) \tag{1.136}$$

In weak-constrained 4dvar the assumption is made that the model errors are Gaussian distributed, so that we can write

$$p(x|y) \propto p(y|x)p(x) \propto \exp(-J(x)) \tag{1.137}$$

in which $J(x)$ is given by:

$$J(x) = \frac{1}{2}\left(x^0 - x_b)\right)^T B^{-1}\left(x^0 - x_b\right)$$

$$+ \sum_{j=1}^{M}\frac{1}{2}\left(y^j - H^j(x^j))\right)^T R^{-1}\left(y^j - H^j(x^j))\right)$$

$$+ \sum_{j=1}^{M}\frac{1}{2}\left(x^j - f(x_i^{j-1})\right)^T Q^{-1}\left(x^j - f(x_i^{j-1})\right) \tag{1.138}$$

in which Q is the covariance of the model errors. This would allow for extension of the window length as the stochastic part of the model equations tend to limit the memory of the system leading to much better convergence of the algorithm. To make the long-window 4Dvar affordable parallelisation is a must. In the following we discuss methods to make the algorithm parallel. Because the 4DVar is an iterative algorithm the parallelisation can only be achieved by splitting up the time window into smaller chunks which are then treated in parallel.

The costfunction can be viewed in two ways: either the unknowns are the states at all time steps, or the unknowns are the state at time zero and all stochastic forcing terms. Let us concentrate on the first view first. To ease the presentation we concentrate on the inner loops linearised around a first-guess model trajectory from a pure model run (with or without stochastic forcing). Denote the latest estimate in the minimisation as $x_{(i)}^k$ with (i) the iteration number, for each time k and write $\delta x^0 = x^0 - x_{(i)}^0$ and $b = x_b - x_{(i)}^0$. We also have

$$x^j - f(x^{j-1}) = x^j - x_{(i)}^j + x_{(i)}^j - f(x^{j-1} - x_{(i)}^{j-1} + x_{(i)}^{j-1}) \approx \delta x^j - F^j \delta x^{j-1} + \beta_{(i)}^j \tag{1.139}$$

in which we defined $\beta_{(i)}^j = x_{(i)}^j - f(-x_{(i)}^{j-1})$, the estimated stochastic term at iteration (i). The innovation can be approximated as

$$H^j(x^j) - y^j = H(\delta x^j + x_{(i)}^j) - y \approx H\delta x - d \tag{1.140}$$

with $d = H(x_{(i)}^j) - y$.

Using this notation the costfunction can be written compactly as:

$$J(\delta x) = (L\delta x - c)^T D^{-1}(L\delta x - c) + (H\delta x - d)^T R^{-1}(H\delta x - d) \qquad (1.141)$$

with $c = (b, \beta^1, \ldots, \beta^n)^T$, and matrices:

$$L = \begin{pmatrix} I & & & & \\ -F^1 & I & & & \\ & -F^2 & I & & \\ & & & \cdots & \\ & & & -F^n & I \end{pmatrix}$$

$$D = \begin{pmatrix} B & & & \\ & Q & & \\ & & Q & \\ & & & \cdots & \\ & & & & Q \end{pmatrix}$$

This formulation can easily be made parallel because the operations $H^j \delta x^j$ and $F^j \delta x^j$ can be computed independently over each sub-interval j. However, the preconditioning:

$$\delta x = L^{-1}(D^{1/2}\chi + c) \qquad (1.142)$$

leads to difficulties as L^{-1} is equivalent to solving equations of the form $L\delta x = g$ for δx, which has to be solved by forward substitution as (see definition of L):

$$\delta x^j = g^j + F^j \delta x^{j-1} \qquad (1.143)$$

which is sequential and cannot be made parallel. Approximating L doesn't work because D has small eigenvalues (large spatial scales have near zero variance) and D appears directly in the Hessian for χ. Hence it is very difficult to find efficient and cheap preconditioners.

The alternative view uses as unknown $\gamma = L\delta x$ leading to cost function

$$J(\gamma) = (\gamma - c)^T D^{-1}(\gamma - c) + (HL^{-1}\gamma - d)^T R^{-1}(HL^{-1}\gamma - d) \qquad (1.144)$$

Also this form cannot be made parallel because we need L^{-1}.

A solution has been found which explores the Saddle Point Formulation. Rewrite the last cost function as:

$$J(\delta x) = (\gamma - c)^T D^{-1}(\gamma - c) + (\delta w - d)^T R^{-1}(\delta w - d) \qquad (1.145)$$

with conditions $\gamma = L\delta x$ and $\delta w = H\delta x$. We can write this as an unconstrained optimisation problem by introducing Lagrange multipliers as:

$$\mathscr{L}(\delta x, \beta, \delta w, \lambda, \mu) = (\gamma - c)^T D^{-1}(\gamma - c) +$$
$$(\delta w - d)^T R^{-1}(\delta w - d) +$$
$$\lambda^T(\gamma - L\delta x) + \mu^T(\delta w - H\delta x) \qquad (1.146)$$

The extremum is found by using calculus of variations leading to:

$$\frac{\partial \mathscr{L}}{\partial \lambda} = 0 \quad => \quad \gamma = L\delta x$$

$$\frac{\partial \mathscr{L}}{\partial \mu} = 0 \quad => \quad \delta w = H\delta x$$

$$\frac{\partial \mathscr{L}}{\partial \gamma} = 0 \quad => \quad D^{-1}(\gamma - b) + \lambda = 0$$

$$\frac{\partial \mathscr{L}}{\partial \delta w} = 0 \quad => \quad R^{-1}(\delta w - d) + \mu = 0$$

$$\frac{\partial \mathscr{L}}{\partial \delta x} = 0 \quad => \quad L^T \lambda + H^T \mu = 0 \qquad (1.147)$$

Elimination γ and δw leads to the system

$$D\lambda + L\delta x = b$$
$$R\mu + H\delta x = d$$
$$L^T \lambda + H^T \mu = 0 \qquad (1.148)$$

There are a few advantages of this algorithm. Firstly, there are no inverse matrices involved so one can work directly with the covariances. Secondly, L^{-1} does not appear, so it is relatively easy to generate a parallel algorithm. Finally, precondition is possible with L^{-1} because that matrix can be approximated without making the preconditioning very dependent on this approximation. A possibility is

$$L^{-1} = 1 + (1 - L) + (1 - l)^2 + \ldots + (1 - L)^{N_x} \qquad (1.149)$$

which can be truncated at some order $p << N_x - 1$.

6.3.2   What does 4Dvar actually calculate?

As mentioned above, 4Dvar calculates the mode of a pdf. In strong-constraint 4DVar this is the mode of the marginal pdf at the beginning of the time window in which

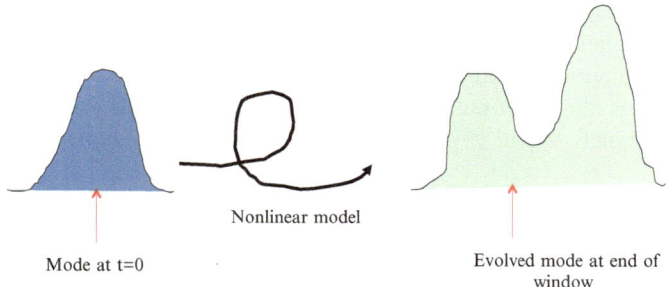

Nonlinear model

Mode at t=0

Evolved mode at end of window

Fig. 7 *Illustration of the problem with strong-constraint 4Dvar: the mode at time zero will not be the mode at the end of the window when the model is nonlinear.*

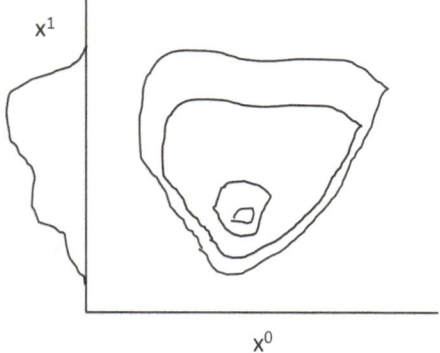

Fig. 8 *Illustration of the problem with weak-constraint 4Dvar: the mode of the joint pdf of the state at two time points $p(x^1, x^0)$ will not be the mode of the marginal pdf $p(x^1)$ at the end of the window when the model is nonlinear.*

we have observations. If the model is linear the evolution of this mode to the end of the window, where the actual forecast begins, leads to the mode of the marginal pdf at the end of the time window. This is, indeed, the best starting position for the forecast.

However, when the model is nonlinear evolving the mode of the pdf at the beginning of the window can lead to a very unlikely state at the end of the time window, see Figure 7. So this could be a very bad initial guess for a forecast. Unfortunately, not much attention to this problem is given in the numerical weather literature.

In weak-constraint 4Dvar the model errors are included and the method calculates the mode of the joint pdf over time. However, also this mode might not lead to the best initial condition for a forecast at the end of the window. The best initial condition for a forecast would be the mode of the marginal pdf at the end of the window. For a nonlinear model the mode of the joint pdf and the mode of the marginal pdf at the end of the window will not be the same, see Figure 8. This is another issue that is largely ignored.

From the above we can conclude that present-day implementations of 4DVar will not provide the best starting point for a forecast, while that is the very reason for its existence in numerical weather prediction. Also, it does not provide an uncertainty estimate, so it is altogether not a good approximation to the posterior pdf. However, it can be extremely useful to generate effective particles, as explained in the next section.

6.3.3   4DVar in a proposal density

The algorithms discussed above cannot be directly used in a particle filter because in a particle filter the particles at time zero, i.e. from the previous posterior pdf, are fixed. So the background term is not part of the cost function. It becomes immediately clear that one has to include model errors to allow these particles to evolve away from the fixed deterministic model, as is consistent with out knowledge of the quality of the models used: model errors are always present in geophysical models, and they tend to be substantial. This means that the now standard strong-constraint 4DVar cannot be used, and we have to rely on the much more limited experience with weak-constraint 4Dvar. This is the reason why we discussed weak-constrained 4Dvar in the previous section. It turns out that leaving out the prior at time zero in the costfunction will still allow us to explore the saddle-point formalism.

The costfunction to be minimised on each particle now becomes:

$$J(x) = \frac{1}{2} \sum_{j=1}^{M} \left(y^j - H_j(x^j) \right)^T R^{-1} \left(y^j - H_j(x^j) \right)$$

$$+ \frac{1}{2} \left(x^n - f(x_i^{n-1}) \right)^T Q^{-1} \left(y^n - f(x_i^{n-1}) \right) \tag{1.150}$$

in which the state at time zero is given by the position of that specific particle at time zero, so from the posterior pdf at time zero.

The 4DVar on each particle is a deterministic solution, and, since the proposal density cannot be a delta function, we have to include a stochastic part somewhere in the algorithm. A natural way to do this is to use the 4DVar in an Implicit Particle Filter. As explained earlier we first draw a random vector from a multivariate Gaussian and then solve for the particle trajectories from:

$$F_i(x^{n-m+1:n}) - \frac{\xi_i^T \xi_i}{2} - \phi_i = 0 \tag{1.151}$$

This equation can be solved efficiently using an iterative Newton method, which will need solving an adjoint equation at each iteration. The connection with 4Dvar is that finding the minimum of the cost function leads to $\partial J/\partial x = 0$, so finding the zero crossing of $\partial J/\partial x$. This is the same problem as finding the zero crossing of (1.151).

The weights will be given by

$$w_i \propto \exp(-\phi_i)\, J_i \tag{1.152}$$

in which J_i is the Jacobian of the transformation, not to be confused with the costfunction. In this we recognise the standard Implicit Particle Filter formulation, as pointed out in Morzfeld et al. (2012) and Atkins et al. (2013).

A special case appears when we assume that the particles at $n - m$ are not fixed but to be drawn from a known $p(x^{n-m})$. In this case we define

$$F(x^{n-m:n}) = -\log(p(y^n|x^n)p(x^{n-m:n})) \tag{1.153}$$

and its minimum

$$\phi = \min(F(x^{n-m:n})) \tag{1.154}$$

where we note that ϕ does not depend on the particle index i. Again we draw a random vector ξ_i of length the size of the state vector times m. The particles are now constructed by solving

$$F(x^{n-m:n}) = \frac{\xi_i^T \xi_i}{2} + \phi \tag{1.155}$$

and the weights become:

$$\begin{aligned}
w_i &= p(y^n|x_i^n)\frac{p(x_i^{n-m:n})}{q_x(x_i^{n-m:n}|y^n)} \\
&= p(y^n|x_i^n)\frac{p(x^{n-m:n})}{q_\xi(\xi_i)}J_i \\
&= \frac{\exp\left(-F_i(x^{n-m:n})\right)}{\exp\left(-1/2\xi^T\xi\right)}J_i \\
&= \frac{\exp(-F_i(x^{n-m:n}))}{\exp\left(-F_i(x^{n-m:n}) + \phi\right)}J_i \\
&\propto J_i
\end{aligned} \tag{1.156}$$

so the weights are only dependent on the Jacobian J_i and not dependent on ϕ_i, leading to much better behaviour of the filter. To use this better behaving scheme we have to be able to draw samples from $p(x^{n-m})$. Doing that we break the sequential nature of the filter, and we have to assume a pdf which is easy to draw from, like a Gaussian.

6.4   The Equivalent-Weights Particle Filter

We have seen that even sophisticated schemes such as the Implicit Particle Filter are likely to show degeneracy when the number of independent observations is large. In this section we discuss a scheme that is not degenerate by construction.

As before we assume we have observations at times $n - m$ and n, and we have an equal weight ensemble of particles at $n - m$, e.g. from a resampling step. We start as in the Implicit Particle Filter with defining:

$$F_i(x^{n-m+1:n}) = -\log(p(y^n|x^n)p(x^{n-m+1:n}|x_i^{n-m})) \qquad (1.157)$$

which is minus the logarithm of the numerator of the final weights. Note that the subscript i refers to the particle position at time $n - m$. Its minimum for each particle is given by

$$\phi_i = \min(F_i(x^{n-m+1:n})) \qquad (1.158)$$

The trick in the Equivalent-Weights particle filter is to set a target weight w_{target} that a set number of particles has to obtain. This target weight relates to a value of ϕ as $\phi_{target} = -\log w_{target}$. Then each particle trajectory is moved in state space between times $n - m$ and n to obtain this weight, so we solve:

$$F_i(x^{n-m+1:n}) = \phi_{target} \qquad (1.159)$$

We can again explore the work-constrained 4DVar algorithm to find a solution to this equation. Call the solution to this equation x_i^*. Solving this equation corresponds to a deterministic move of the particle. Of course, the movement of a particle cannot be fully deterministic. A deterministic proposal would mean that the proposal transition density q is a delta function, and as it appears in the denominator of the weights this leads to division by zero. So we add a stochastic term to the deterministic solution, leading to:

$$x_i^{n-m+1:n} = x_i^* + \beta_i^{n-m+1:n} \qquad (1.160)$$

If the amplitude of β is small, the stochastic move will be small, and the final weight of the particle will not change much from the target weight, leading to particles with equivalent weights.

This scheme has three choices built in that we will discuss below. A first question is how to choose the target weight. Figure 9 contains a schematic that explains the procedure. Firstly, we cannot choose the target weight as the maximum weight that any particle can reach, as only one particle can reach that weight. No matter how the other particles are moved in state space at times $n - m + 1 : n$ their weight will always be smaller. Secondly, choosing the target weight as the maximum weight of the worst particle is also not a good idea. This would mean that most particles

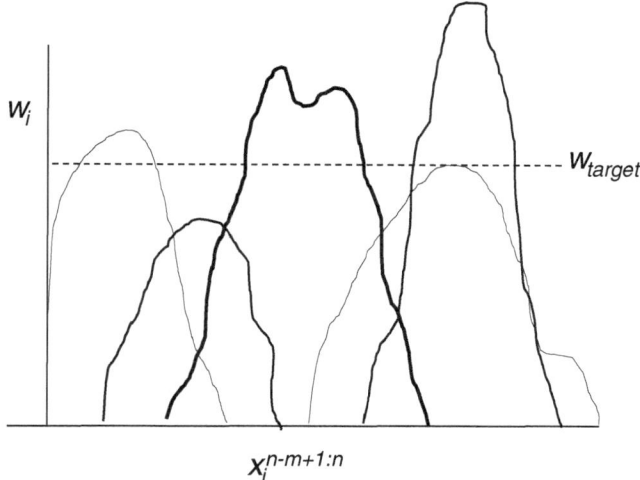

Fig. 9 *Schematic of the weights for 5 particles as function of the position of their trajectory through state space. The target weight is set such that 80% of the particles can reach it.*

will have to move quite far from their optimal value for the weight, so away from the observations, and away from where the deterministic model would like to push them. The best choice is somewhere in between. In our experience a target weight that 80% of the particles can reach works well.

This means that we have to move 80% of the particles such that their weight becomes equal to the target weight. Figure 9 shows a 1-dimensional example in which there are typically 2 solutions for each particle in the 80%. However, when the system is 2-dimensional the solution will lie on the intersection of a plane, defining the target weight, and the shape of the weight for each particle, which is a circular shape, so there are an infinite number of solutions. The question then becomes which solution to choose, and how to find it. We can borrow ideas from the Implicit Particle Filter literature to solve this problem, i.e. using the random map.

And what should one do with the other 20%? They cannot reach the target weight so the above procedure does not apply to them. In fact, since their weight will be much lower than that of the other particles when the number of independent observations is large, we can ignore their evolution completely and resample them form the other 80%.

The final choice to make is on the shape of the proposal density, so the density from which final stochastic move is drawn. The proposal transition density could be chosen a Gaussian, but since the weights have q in the denominator a draw from the tail of a Gaussian would lead to a very high weight for a particle that is perturbed by a relatively large amount. To avoid this q can be chosen as a mixture density

$$q(x_i^{n-m+1:n}|x_i^{n-m}) = (1 - \gamma)U(-a,a) + \gamma N(0, a^2) \qquad (1.161)$$

in which γ and a are small. By choosing γ small the change of having to choose from $N(0, a^2)$ can be made as small as desired. For instance, it can be made dependent on the number of particles N. a has to be small to ensure the stochastic movement is small so that the weight of the particles will not change much and this close to the target weight.

We can study how the scheme behaves for linear models and linear observation operators H. In that case we can write, as for the implicit particle filter:

$$F_i(x^{n-m+1:n}) = (x_i^{n-m+1:n} - x_i^a)^T P^{-1} (x_i^{n-m+1:n} - x_i^a) + \phi_i \tag{1.162}$$

so we find:

$$(x_i^{n-m+1:n} - x_i^a)^T P^{-1} (x_i^{n-m+1:n} - x_i^a) + \phi_i = \phi_{target} \tag{1.163}$$

A possible solution to this equation can be written as:

$$x_i^a = Mx_i^{n-m} + \alpha_i K(y - H^n Mx_i^{n-m}) \tag{1.164}$$

in which α_i is a scalar, and, as before, Mx_i^{n-m} generates a trajectory $(x_i^{n-m+1}, \ldots, x_i^n)$ using the deterministic model. Clearly, if $\alpha = 1$ we find the solution as proposed in the Implicit Particle Filter. We choose the scalar α_i such that the weights are equal, so we use this expression in equation (1.163), leading to

$$\alpha = 1 - \sqrt{1 - b_i/a_i} \tag{1.165}$$

in which $a_i = 0.5 d_i^T R^{-1} H^n K d$ and $b_i = 0.5 d_i^T R^{-1} d_i - \phi_{target}$. Here $d_i = y^n - H^n Mx_i^{n-m}$.

It is good to realise what we have achieved: a particle filter that generates samples from the high-probability area of the posterior pdf which is not degenerate by construction. The only drawback is that the scheme has one tuning parameter, the percentage of particles kept within the scheme. Experience with a simplified version of the scheme explored below shows that the quality of the method does depend on this choice.

6.4.1 Convergence of the EWPF

A simple requirement for any particle filter seems to be that it should converge to the full posterior from Bayes Theorem when the number of particles increases. However, it is easy to show that this is not the case for the Equivalent-Weights Particle Filter. The main issue is the percentage of particles kept in the equal-weights step. Whatever percentage is chosen we will always lose the tails of the density. Furthermore, a very high percentage will push all particles away from their highest weight, pushing them towards the tails of the posterior (but also never pushing far enough into the tails to cover them because the percentage will always be finite).

There is, however, a very simple remedy. One can make the proposal moves converge to that of the original model with increasing number of particles by, e.g., multiplying the deterministic moves away from the original deterministic part of the model by a factor that decreases these terms to zero when N increases. For instance, we can multiply α_i is the equivalent-weights step by a factor $1000/(1000+N)$. It is possible to tune this such as to obtain a smooth transition between the EWPF when N is small, and the SIR when N grows larger.

However, I don't consider this a fruitful way. The EWPF generates high-probability particles from the posterior pdf when N is small. Experiments shown above and in, e.g., Ades and Van Leeuwen (2012); Ades and van Leeuwen (2014) show that the method is very effective doing so, and the ensemble statistics are good, even for very high dimensional systems.

6.4.2   Simple implementations for high-dimensional systems

So far the Equivalent-Weights Particle Filter has been tested on several high-dimensional geophysical systems, but only in a form in which the actual equivalent-weights step is used at the final step before observations. This implementation avoids complicated iterative searches over several time steps. Between observations a relaxation scheme is used to ensure the particles are reasonably positioned before the Equivalent-Weights step. At the beginning of this chapter we discussed a very simple relaxation scheme to pull the particles towards future observations, written as:

$$x_i^j = f(x_i^{j-1}) + \beta_i^j + T^j(y^n - H(x_i^{j-1})) \tag{1.166}$$

in which n is the next observation time. Unfortunately, this proposal density will not avoid degeneracy in high-dimensional systems with a large number of observations, so we will need an equivalent-weights step at the final time step.

To start, let us recall the expression we found for the proposal density weights in equation 1.64:

$$w_i = p(y^n|x_i^n) \prod_{j=n-m+1}^{n} \frac{p(x_i^j|x_i^{j-1})}{q(x_i^j|x_i^{j-1}, y^n)} \tag{1.167}$$

The model with the extra relaxation term will be used for all but the last time step before observations. This last step will employ the equivalent-weights step as explained in the previous section: first perform a deterministic time step with each particle that ensures that most of the particles have equal weight, and then add a very small random step to ensure that Bayes theorem is satisfied, see also Van Leeuwen (2010, 2011) for details. For the first stage we write down the weight for each particle using only a deterministic move, so ignoring the proposal density q for the moment:

$$-\log w_i = F_i(x^n) \tag{1.168}$$

which for Gaussian distributed model errors and observation errors reduces to

$$F_i(x^n) = \log w_i^{rest} + \frac{1}{2}(y^n - Hx_i^n)^T R^{-1}(y^n - Hx_i^n) + \frac{1}{2}(x_i^n - f(x_i^{n-1}))^T Q^{-1}(x_i^n - f(x_i^{n-1}))$$

(1.169)

in which w_i^{rest} is the weight accumulated over the previous time steps between observations, so the p/q factors from each time step.

If H is linear, which is not essential but as we will assume for simplicity here, this is a quadratic equation in the unknown x_i^n. All other quantities are given. We calculate the minimum of this function for each particle i, which is simply given by

$$\phi_i = -\log w_i^{rest} + \frac{1}{2}\left(y^n - Hf(x_i^{n-1})\right)^T \left(HQH^T + R\right)^{-1}\left(y^n - Hf(x_i^{n-1})\right) \quad (1.170)$$

For N particles this given rise to N minima. Next, we determine a target weight as the weight that 80% of the particles can reach, i.e. 80% of the minimum ϕ_i is smaller than the target value. (Note that we can choose another percentage, see, e.g., Ades and Van Leeuwen (2012) who investigate the sensitivity of the filter for values between 70% and 100%.) Define a quantity $C = -\log w_{target}$, and solve for each particle with a minimum weight larger than the target weight

$$F_i(x^n) = -\log(w_{target}) \tag{1.171}$$

So now we have found the positions of the new particles x_i^* such that all 80% have equal weight. The particles that have a larger minimum than $-\log w_{target}$ will come back into the ensemble via a resampling step, to be discussed later.

The equation above has an infinite number of solutions for dimensions larger than 1. To make a choice we assume

$$x_i^n = f(x_i^{n-1}) + \alpha_i K[y^n - Hf(x_i^{n-1})] \tag{1.172}$$

in which $K = QH^T(HQH^T + R)^{-1}$, Q is the error covariance of the model errors, and R is the error covariance of the observations. Clearly, if $\alpha_i = 1$ we find the minimum back. We choose the scalar α_i such that the weights are equal, leading to

$$\alpha = 1 - \sqrt{1 - b_i/a_i} \tag{1.173}$$

in which $a_i = 0.5x_i^T R^{-1}HKx$ and $b_i = 0.5x_i^T R^{-1}x_i - \log w_{target} - \log w_i^{rest}$. Here $x = y^n - Hf(x_i^{n-1})$, and w_i^{rest} denotes the relative weights of each particle i up to this time step, related to the proposal density explained above.

Of course, this last step towards the observations cannot be fully deterministic and as explained in the previous section we choose a mixture density:

$$q(x_i^n|x^*) = (1 - \gamma)U(-a, a) + \gamma N(0, a^2) \tag{1.174}$$

in which x_i* the particle before the last random step, and γ and a are small. By choosing γ small the change of having to choose from $N(0, a^2)$ can be made as small as desired. For instance, it can be made dependent on the number of particles N.

To conclude, the equivalent-weight scheme explored in several applications so far consists of the following steps:

1 Use the modified model equations for each particle for all time steps between observations.
2 Calculate for each particle i for each of these time steps

$$w_i^j = w_i^{j-1} \frac{p(x_i^j|x_i^{j-1})}{q(x_i^j|x_i^{j-1}, y^n)} \tag{1.175}$$

3 At the last time step before the observations calculate the maximum weights for each particles and determine $F_i(x^n) = -\log w_{target}$
4 Determine the deterministic moves by solving for α_i for each particle as outlined above.
5 Choose a random move for each particle from the proposal density (1.161).
6 Add these random move to each deterministic move, and calculate the full posterior weight.
7 Resample, and include the particles that have been neglected from step 4 on.

It is stressed again that we do solve the fully nonlinear data assimilation problem with this efficient particle filter, and the only approximation is in the ensemble size. All other steps are completely compensated for in Bayes Theorem via the proposal density freedom, although the performance of the scheme is dependent on the correct tuning of the scheme parameters T^j and the percentage of particles kept in the equivalent weights step. This method has been studied quite extensively in low-dimensional models (Ades and Van Leeuwen, 2012) and used successfully in several high-dimensional applications, see, e.g., Van Leeuwen and Ades (2013), Ades and van Leeuwen (2014) for an application in a 65,000 dimensional barotropic vorticity equation. We are testing its performance in even higher-dimensional climate models with millions degrees of freedom at the moment.

To illustrate its performance we follow Ades and van Leeuwen (2014) by showing some insightful extra results. Ades and van Leeuwen (2014) applied the method to the so-called barotropic vorticity equation model, with a dimension of 65,536. The model solves the evolution equation for the vorticity q:

$$\frac{\partial q}{\partial t} + J(\psi, q) = \beta \tag{1.176}$$

in which ψ is the streamfunction, related to the vorticity as $q = \Delta\psi$, and Δ is the 2-dimensional Laplace operator. The Jacobian $J(a, b) = a_x b_y - a_y b_x$ in which subscripts denote derivatives to the zonal and meridional coordinates x and y, respectively. β is a random space and time dependent forcing representing missing physics. The equations are discretised over a equidistant horizontal grid on a torus, with 256×256 grid points. The evolution equation is solved with a semi-Lagrangian scheme with a time step $\Delta t = 0.04$. The time stepper itself was Euler-Maruyama. After each time step the stream function ψ is calculated using FFTs. This stream function field is then used in the next time step to advect the vorticity field further.

This system describes the nonlinear evolution of interacting vortices on a torus. The data assimilation experiment is a so-called identical twin experiment, in which first a truth run is generated using the stochastic model equations and observations are taken from this truth run. These artificial observations are then perturbed by the prescribed observation errors and then assimilated into model runs with different initial condition and stochastic forcing realisations, but using the same model dynamics and statistics for model errors and observation errors.

We compare the results of a data experiment using the SIR filter, which only uses resampling, and the Equivalent-Weights Particle Filter, both with 32 particles. 80% of the particles was kept.

In our case the experiment was run for 1150 time steps with observations every 50 time steps. As the decorrelation timescale, determined from a single model run as the time scale at which the autocorrelation drops below $1/e$, is 42 time steps, the system has time to develop nonlinearly between observation times. The observations are assumed to be uncorrelated in space and time, with standard deviation 0.05. In this experiment we observe every grid point at observation time.

The initial condition was a random field of vortices with length scales of 10 grid points. The initial ensemble was generated by adding Gaussian random fields with variance 0.025^2 and spatial correlations given by

$$\rho(d) = \left(1 + \frac{|d|}{L}\right) \exp\left(-\frac{|d|}{L}\right) \tag{1.177}$$

in which d is the Eulerian distance between grid points and L is the decorrelation length scale of 5 grid points. This correlation structure was also used for the model error covariance, but now with variance $0.025^2 \Delta t$. This random error gives perturbations to the deterministic model move of about 10%.

Figure 10 shows the truth, the mean of the ensemble using the SIR filter that only used resampling, and the EWPF. We clearly see that the EWPF manages to capture the position and strength of the vortices, but also the majority of the filament structure. In contrast, the SIR result is strongly different from the truth. This filter is indeed degenerate from the first observation set at day 50 on, as expected.

Figure 11 explores the quality of the ensemble spread. It shows the squared error $(mean - truth)^2$ for the EWPF at time 600 and time 1150, compared to the variance in the ensemble. If the filter does well, these two should be quite similar. It should be noted that the variance is a statistical measure, while the $(mean - truth)^2$ is a realisation, so the latter is expected to be much more variable, as indeed shown in the figure. Note that the structures in squared error and variance are similar. Also note that the spatial average is 0.046 versus 0.031 at time 600, and 0.0032 versus 0.0030 at time 1150, showing that the squared error and the variance are similar in magnitude.

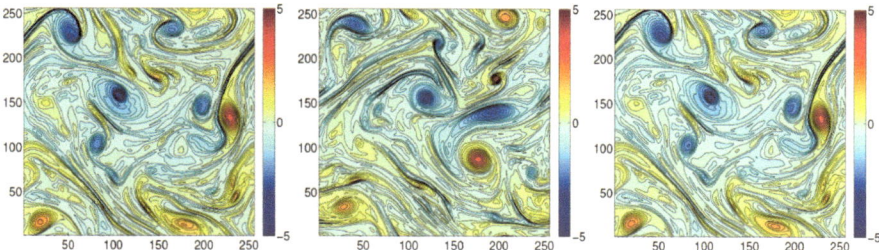

Fig. 10 *Vorticity field of the truth (left), the mean of the SIR filter (middle) and the mean of the EWPF (right) at time 1150. Note that the EWPF is quite close to the truth, while the SIR is completely off.*

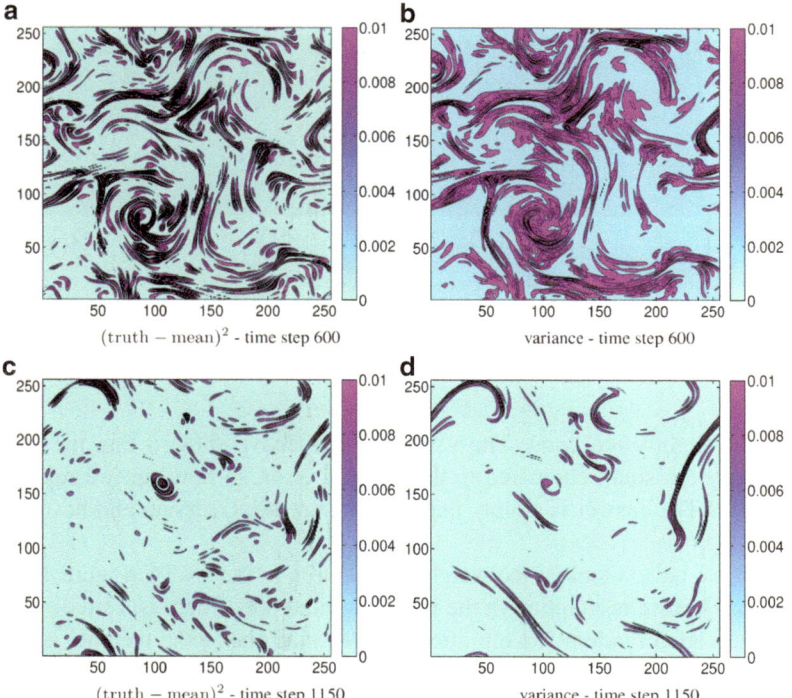

Fig. 11 *Comparison of the squared error (mean − truth)² with the variance in the ensemble at time 600 and at time 1150 for the EWPF. Note that the structures in error and variance are similar and that the size is also similar. See text for a more detailed discussion.*

To further analyse the quality of the ensemble we calculated the rank histogram. This histogram is constructed by ranking the truth in the ensemble for a collection of grid points at all observation times. Figure 12 shows that the rank histogram is slightly bulging, meaning that the truth ranks more in the middle of the ensemble than at the extremes. This shows that the ensemble is slightly over dispersive:

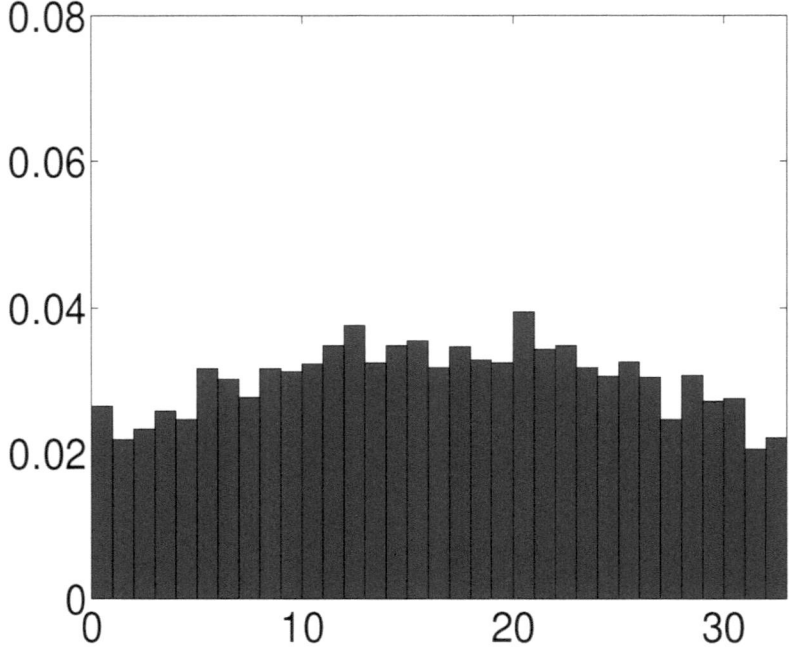

Fig. 12 *Rank histogram of the ranking of the truth in the ensemble, averaged over several grid points and over observation times. The non-flat shape illustrates that the ensemble is slightly overdispersive.*

it's spread slightly too wide. The variance plots above showed that the variance was slightly too small compared to the squared error, so how can we relate these two results? The answer is simply that the pdfs are not Gaussian and have heavier tails.

At last we have a look at the weight distribution in the EWPF. Figure 13 shows the weights before resampling in the EWPF at time 600. There is a small variation due to the added random forcing after the step that makes the weights equal, but apart from that 25 particles (80 %) have very similar weights, showing that the filter is not degenerate for this high-dimensional example. This is, of course, due to the construction of the filter. Note that 7 particles weights have a much lower weight; they come back via resampling.

6.4.3 Comparison of nonlinear data assimilation methods

In this section we compare the standard particle filter (SIR), Equivalent-Weights Particle Filter (EWPF) and the Implicit Particle Filter (IPF) with the following Metropolis-Hastings based methods: the random walk Metropolis Hastings (MH), the preconditioned Crank-Nicholson Metropolis-Hastings (MHpCN), and Metropolis-adjusted Langevin (MALA). We use a simple model given by:

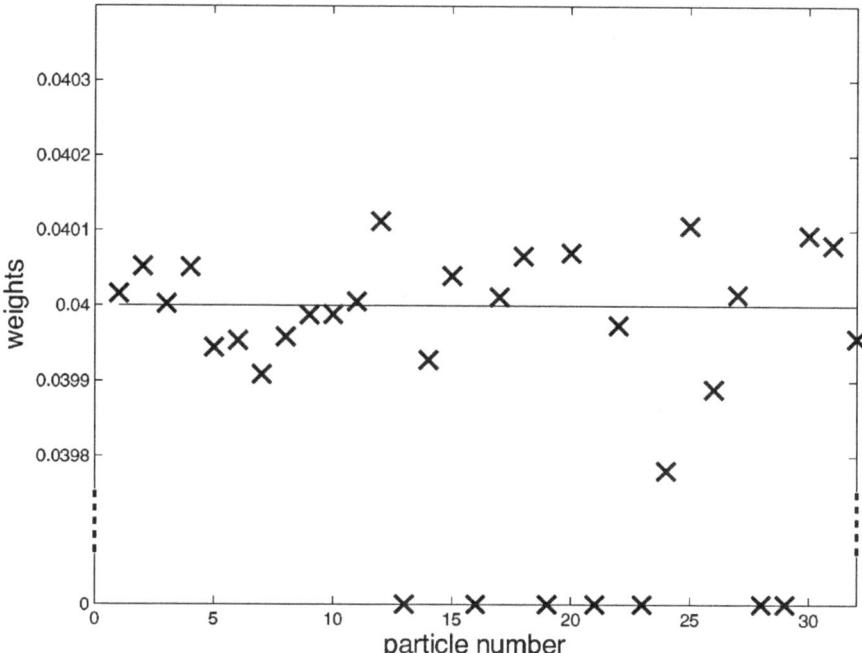

Fig. 13 *Weights of the particles before resampling. Note that the filter is not degenerate by construction as 25 particles have a good weight. The other 7 particles come back into the ensemble via resampling.*

$$x^1 = x^0 + \eta \tag{1.178}$$

The distribution of x^0 is taken as independent Gaussian for simplicity. The state vector is N_x-dimensional and we will investigate the performance of the filters when N_x increases. Finally, the model errors are also independent Gaussian. The state x^1 is observed as

$$y = x^1 + \epsilon \tag{1.179}$$

in which the observation errors are also taken independent and Gaussian.

While this system is completely Gaussian, so linear, so the traditional linear data-assimilation methods can be used, we will explore several nonlinear data-assimilation methods here. The idea is that if the methods fail in this simple linear case it is unlikely they will perform better in a nonlinear setting.

Several experiments were performed with different values for initial, model, and observational errors. Here we report on the following experimental settings, mentioning that other settings give similar results: the initial mean is 0 for each variable, the initial variance is 1 for each variable, the model variance is 0.01, and the observation error variance is 0.16. The number of ensemble members or particles

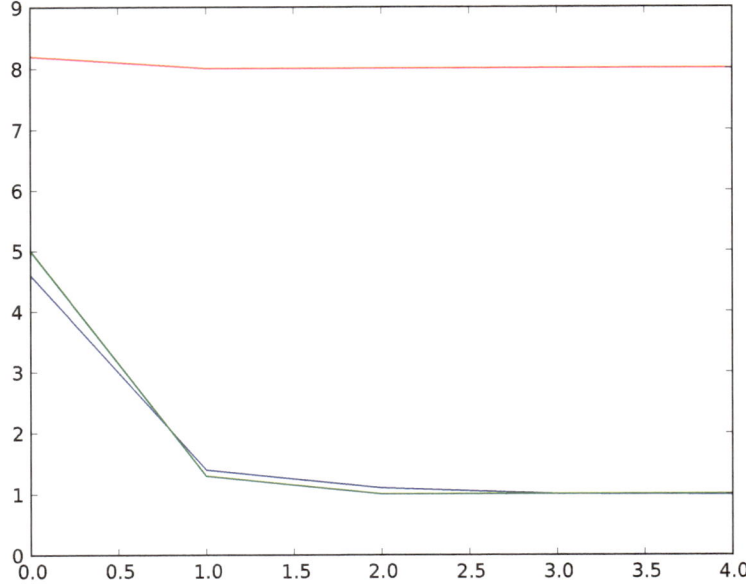

Fig. 14 *Effective ensemble size for SIR (blue), EWPF (red), and IPF (green) as function of the power of the state dimension, so* 10^0 *to* 10^4.

or MCMC samples is 10 in all experiments. For the equivalent-weights we keep 80% of the particles. For the Metropolis-Hastings sampler we set the width of the Gaussian proposal such that the acceptance rate is 30–50%. This let to the lowest root-mean-square error (RMSE). For Metropolis-Hastings with the preconditioned Crank-Nicholson proposal we scale the factor γ such that the acceptance rate is 30–50%. This let to the lowest RMSE. Finally, for Metropolis-adjusted Langevin we set the step size δ such that the acceptance rate was 30–50%, again leading to best performance.

Figure 14 shows the effective ensemble size in the three particle filters tested, defined as

$$N_{eff} = \frac{1}{\sum w_i^2} \qquad (1.180)$$

for the standard particle filter with proposal density equal to the prior (SIR), the Equivalent-Weights Particle Filter (EWPF) and the Implicit Particle Filter (IPF), which is equal to a particle filter using the so-called Optimal Proposal density. The results shown are averages over 100 experiments. We can clearly see that apart from a state dimension of 1, all filters are degenerate, except for the EWPF, in which we find constant effective ensemble sizes of 8 out of 10, identical to the percentage of particles kept in the equivalent-weights procedure.

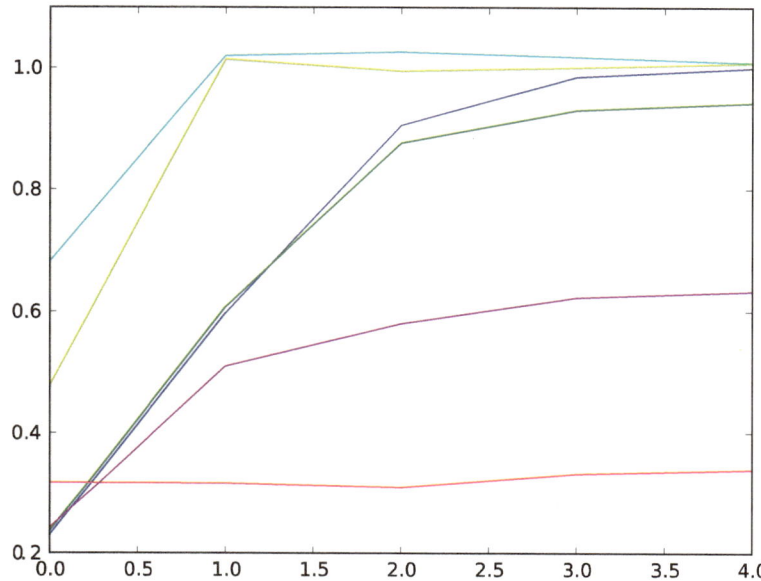

Fig. 15 *State-space averaged root-mean-square error of ensemble mean for SIR (blue), EWPF (red), IPF (green), MH (light blue), MHpCN (purple), MALA (light green) as function of the power of the state dimension, so 10^0 to 10^4.*

Figure 15 shows the root-mean-square error (RMSE) of the ensemble mean from the truth for each method, averaged over state space. The results shown are again averages over 100 experiments.

The theoretical posterior variance per dimension is 0.37 for this case, so a good filter would have an RMSE close to that. For dimension 1 we observe that most methods do reach that value, apart from MH, which is related to the small sample size. However, when the state dimension increases to 10 or higher only the EWPF keeps a proper RMSE, all other methods get a way too high value, close to the prior estimate. Of note is the MHpCN which settles around a value of 0.6. It is unclear why this method outperforms the others in this way (apart from the EWPF).

One might wonder why the IPF does not do much better than the standard particle filter. The reason is simply that the members are moved to better positions but the filter is degenerate, so the mean only consist of the best particle, which does depend on the initial particle position, so the RMSE will converge to that of the prior. Note that, for this specific case in which the prior at time zero is a Gaussian one could take the drawing from that Gaussian into the sampling via the proposal q, in which case all samples would have equal weight as this system is linear. In that case the IPF reduces to the Ensemble Smoother of Van Leeuwen and Evensen (1996, or the Ensemble Kalman Smoother of Evensen and Van Leeuwen, 2000). The point here is that, in general, the prior at time zero is non-Gaussian as it arises from a sequential application of the algorithm so the starting point at time zero is a number of particles with unknown distribution.

The results here do depend on the specific choice of the system parameters, i.e. the model error covariance, the observation covariance, and the covariance of the initial ensemble. For instance, the IPF has a much lower RMSE (0.45 for $N_x =$ 10) when the model variance is multiplied by a factor 50. However, the effective ensemble size remains below 2, even for a 10-dimensional system, showing that the filter cannot represent the pdf with a low number of particles, which is the goal in geophysical data assimilation. These findings are consistent with our order of magnitude estimates on the optimal proposal density earlier. The point is, however, that even when varying these system parameters the general conclusion remains that only the EWPF is able to keep the effective number of particles high and the RMSE low.

7   Conclusions

The nonlinear data-assimilation problem for high-dimensional systems is a hard one and this paper tries to summarise what has been done, providing new links and a few new ideas. It was shown that data-assimilation is not an inverse but a multiplication problem, and the goal of data assimilation is to try to represent the posterior pdf as efficiently as possible. That does not mean, however, that inverse methods are not important for the data-assimilation problem. The vast literature on inverse methods has resulted a huge variety of methods that can be, and are, explored in the more advanced Markov-Chain Monte-Carlo methods and in Particle Filters. Examples discussed here include so-called variational methods, but there is much more to explore, like iterative Tikhonov regularisation. Of specific mention is are variants that can handle non-smooth problems, see, e.g., Steward et al. (2012) for a geophysical example. All these methods can be part of the proposal densities in the Bayesian framework, and this seems to be the natural way to combine the Bayesian and the inverse problems fields, and indeed their communities. In nonlinear systems the pdf is typically not of a shape that can be described by a few parameters, so one typically resorts to representing the posterior pdf by a set of samples from that pdf. This then automatically leads to Monte-Carlo methods to solve the nonlinear data-assimilation problem.

Although there is a vast variety of nonlinear data-assimilation methods available, only a few are actually applicable to high-dimensional problems. The problems in mind are of geophysical origin, in which the state dimension can easily exceed one million. We discussed the applicability of Markov-Chain based methods like the Gibbs Sampler, Metropolis-Hastings, Hybrid Monte-Carlo and (Metropolis-adjusted) Langevin Monte-Carlo. All these methods rely on the notion of a Markov Chain that is used to generate the next sample from the previous sample. This immediately points to a problem with these methods: the next sample is typically not independent of the previous sample, so several subsequent samples have to be discarded before a true new independent sample from the posterior is found. This makes these methods less efficient from the start.

Particle filters are among the few that do not have this problem. They have a few strong assets, i.e. their full nonlinearity, the simplicity of their implementation (although this tends to be lost in more advanced variants), the fact that balances are automatically fulfilled (although, again, more advanced methods might break this), and, quite importantly, that their behaviour does not depend on a correct specification of the model state covariance matrix.

We have also seen the weaknesses in terms of efficiency, the filter degeneracy problem, that plagues the simpler implementations. However, recent progress seems to suggest that we are quite close to solving this problem with developments like the Implicit Particle Filter and the Equivalent-Weights Particle Filter. Interestingly, by exploring proposal densities we are effectively using some sort of localisation as each grid point typically only sees observations within an influence region set by the covariance matrix used in the proposal. This does reduce the degeneracy problems considerably. Direct localisation can also now be explored by using ideas from optimal transportation. Strong new developments are expected when exploring these ideas further.

There is a wealth of new approximate particle filters that typically shift between a full particle filter and an ensemble Kalman filter, depending on the degeneracy encountered. Especially Gaussian mixture models for the prior are popular. I have refrained from trying to give an overview here, there is just too much going on in this area. A brief discussion is given in Van Leeuwen (2009), but that is largely out of date now. In a few years time we will have learned what is useful and what is not.

Finally, it must be said that the methods discussed above have a strong bias to state estimation. One could argue that this is fine for prediction purposes, but for model improvement (and thus indirectly forecasting) parameter estimation is of more interest (and what about parameterisation estimation...). Unfortunately no efficient Particle Filter schemes exist for that problem. This is a growing field that needs much more input from bright scientists.

Acknowledgements I thank the members Data Assimilation Research Centre (DARC) for numerous discussions on the topics of this paper, especially Melanie Ades, Javier Amezcua, David Livings, Phil Browne and Sanita Carvalho-Vetra. None of them is in anyway responsible for the contents. I also thank funding from the National Environment Research Council (NERC) via the National Centre of Earth Observation (NCEO) and via several other grants, and the EU FP7 project SANGOMA.

References

Ades, M., and P.J. Van Leeuwen, An exploration of the Equivalent-Weights Particle Filter, Q. J. Royal Meteorol. Soc., doi:10.1002/qj.1995, 2012.

Ades, M., and P.J. van Leeuwen, The equivalent-weights Particle Filter in a high-dimensional system, Monthly Weather Rev., accepted, 2014. Chorin A.J., X. Tu Implicit sampling for particle filters. *PNAS* 106, 17249–17254. 2009.

Atkins, E., M. Morzfeld, X. Tu, and A.J. Chorin, Implicit particle methods and their connection with variational data assimilation, Monthly Weather Review 141, 1786–1803, 2013.

Beskos, A., G. Roberts, A. Stuart, and J. Vos, MCMC methods for diffusion bridges, Stoch. Dyn. 8, 319–350, 2008.

Beskos, A., Pinski, F.J., Sanz-Serna, J.M. and Stuart, A.M.. Hybrid Monte Carlo on Hilbert spaces. Stochastic Process. Appl. 121 2201–2230. MR2822774, 2011.

Beskos, A., N. Pillai, G. Roberts, J.-M. Sanz-Serna and A.M. Stuart, Optimal tuning of the hybrid Monte Carlo algorithm. Bernoulli 19, 1501–1534, 2013.

Beyou, S., A. Cuzol, S. Subrahmanyam Gorthi, and E. Mémin. Weighted ensemble transform Kalman filter for image assimilation. Tellus A, 65, 2013.

Burgers, G., P.J. van Leeuwen and G. Evensen. Analysis Scheme in the Ensemble Kalman Filter. Monthly Weather Rev., 126, 1719–1724, 1998.

Chorin A.J., X. Tu. Implicit sampling for particle filters. PNAS 106, 17249–17254, 2009.

Chorin A.J., Morzfeld M. and Tu X. Interpolation and iteration for nonlinear filters. Commun. Appl. Math. Comput. Sci. 5, 221–240, 2010.

Cohn. An Introduction to Estimation Theory. J. Met. Soc. Japan, 75(1B), 1997.

Cotter, S.L., G.O. Roberts, A.M. Stuart, and D. White, MCMC methods for functions: Modifying old algorithms to make them faster, Stat. Science, 28, 424–446, DOI:10.1214/13-STS421, 2013.

Doucet, A., N. De Freitas, and N. Gordon, *Sequential Monte-Carlo methods in practice*, Springer-Verlag, Berlin, p. 581, 2001.

Duane, S., A.D. Kennedy, B. Pendleton, and D. Roweth, Hybrid Monte-Carlo, Phys. Lett. B, 195, 216–222, 1987.

Evensen, G. Data Assimilation: The Ensemble Kalman Filter, Springer, 2009.

Evensen G. Sequential data assimilation with a nonlinear quasi-geostrophic model using Monte-Carlo methods to forecast error statistics. J. Geophys. Res., 99, 10143–10162, 1994.

Evensen, G. and P.J. van Leeuwen. An Ensemble Kalman Smoother for nonlinear dynamics. Monthly Weather Rev. 128, 1852–1867, 2000.

Fisher, M., and H. Auvinen, Long window 4DVar, Proceedings of the ECMWF Seminar on Data Assimilation for Atmosphere and Ocean, 6–9 September 2011, 2012.

Gordon, N.J., D.J. Salmond, and A.F.M. Smith., "Novel approach to nonlinear/non-Gaussian Bayesian state estimation", *IEE Proceedings-F*, 140, 107–113, 1993.

Gu Y, Oliver DS, An iterative ensemble Kalman filter for multiphase fluid flow data assimilation. SPE J 12(4):438–446, 2007.

Huber, G., 1982: Gamma function derivation of n-sphere volumes. American Mathematical Monthly, 89, 301–302.

Houtekamer, P.L. and H.L. Mitchell, Data assimilation using an ensemble Kalman Flter technique, Mon. Wea. Rev., 126, 796–811, 1998.

Jazwinski, A.H. Stochastic Filtering and Filtering Theory, Academic Press, 1970.

Kim, S., G.L. Eyink, J.M. Restrepo, F.J. Alexander, and G. Johnson, 2003: Ensemble filtering for nonlinear dynamics, *Monthly Weather Rev.*, **131**, 2586–2594.

Le Gland, F., V. Monbet and V-D. Tran, Large sample asymptotics for the ensemble Kalman filter, in Handbook on Nonlinear Filtering, D. Crisan and B. Rozovskii, editors, pp. 598–631, Oxford University Press, Oxford, 2011.

Morzfeld, M., X. Tu, E. Atkins, and A.J. Chorin, A random map implementation of implicit filters, Journal of Computational Physics 231, 2049–2066, 2012.

Papadakis N., E. Mémin, A. Cuzol, and N. Gengembre. Data assimilation with the weighted ensemble Kalman filter. Tellus, 62A:673–697, 2010.

Pitt, M.K., and N. Shephard, 1999: Filtering via simulation: Auxillary particle filters, *J. American Stat. Ass.*, **94**, 590–599.

S. Reich, A non-parametric ensemble transform method for Bayesian inference. SIAM J Sci Comput, 35 , A2013-A2014, 2013.

Robert, Ch.P., and G. Casella, Monte-Carlo Statistical Methods, Springer, 2004.

Snyder, C., T. Bengtsson, P. Bickel, and J. Anderson "Obstacles to high-dimensional particle filtering", *Mon. Wea. Rev.*, 136, 4629–4640, 2008.

Steward, J.L., I.M. Navon, M. Zupanski and N. Karmitsa. Impact of Non-Smooth Observation Operators on Variational and Sequential Data Assimilation for a Limited-Area Shallow Water Equations Model. Quart. Jour. Roy. Met Soc. , 138, 323–339, 2012.

M.K. Tippett, J.L. Anderson, C.H. Bishop, T.M. Hamill and J.S. Whitaker. Ensemble square root filters. Monthly Weather Review, 131, 1485–1490, 2003.

Van Leeuwen, P.J. and G. Evensen. Data assimilation and inverse methods in terms of a probabilistic formulation. Monthly Weather Rev. 124, 2898–2913, 1996.

Van Leeuwen, P.J. Particle Filtering in Geosciences, *Monthly Weather Rev.*, 137, 4089–4114, 2009.

Van Leeuwen, P.J., Nonlinear Data Assimilation in geosciences: an extremely efficient particle filter, *Quart. J. Roy. Meteor. Soc.*, 136, 1991–1996, 2010.

Van Leeuwen, P.J.: Efficient non-linear Data Assimilation in Geophysical Fluid Dynamics *Computers and Fluids*, doi:10.1016/j.compfluid.2010.11.011, 2011.

Van Leeuwen PJ and Ades M. Efficient fully nonlinear data assimilation for geophysical fluid dynamics. Comput. Geosci. 55: 16–27, doi: 10.1016/j.cageo.2012.04.015, 2013.

Van Leeuwen, P.J.: Representation errors in Data Assimilation, Q.J.R. Meteorol. Soc., 2014, DOI: 10.1002/qj.2464

Zupanski, M., Maximum Likelihood Ensemble Filter: Theoretical aspects, Mon. Weather Rev., 133, 1710–1726, 2005.

Chapter 2
Assimilating data into scientific models: An optimal coupling perspective

Yuan Cheng and Sebastian Reich

Abstract Data assimilation is the task of combining mathematical models with observational data. From a mathematical perspective data assimilation leads to Bayesian inference problems which can be formulated in terms of Feynman-Kac formulae. In this paper we focus on the sequential nature of many data assimilation problems and their numerical implementation in form of Monte Carlo methods. We demonstrate how sequential data assimilation can be interpreted in terms of time-dependent dynamical systems or, more generally, time-dependent Markov processes, which is often referred to as the McKean approach to Feynman-Kac formulae. It is shown that the McKean approach has very natural links to coupling of probability measures and optimal transportation. This link allows one to propose novel sequential Monte Carlo methods/particle filters. In combination with localization these novel algorithms have the potential of beating the curse of dimensionality, which has prevented particle filters from being applied to spatially extended systems.

1 Introduction

This paper is concerned with Monte Carlo methods for approximating expectation values for sequences of probability density functions (PDFs) $\pi^n(z)$, $n \geq 0$, $z \in \mathscr{Z}$. We assume that these PDFs arise sequentially from a Markov process with given transition kernel $\pi(z|z')$ and are modified by weight functions $G^n(z) \geq 0$ at each iteration index $n \geq 1$. More precisely, the PDFs satisfy the recursion

Y. Cheng
Institut für Mathematik, Universität Potsdam, Am Neuen Palais 10, D-14469 Potsdam, Germany
e-mail: yuan.cheng@uni-potsdam.de

S. Reich (✉)
Institut für Mathematik, Universität Potsdam, Am Neuen Palais 10, D-14469 Potsdam, Germany

Department of Mathematics and Statistics, University of Reading, Whiteknights,
PO Box 220, Reading RG6 6AX, UK
e-mail: sreich@math.uni-potsdam.de

© Springer International Publishing Switzerland 2015
P.J. van Leeuwen et al., *Nonlinear Data Assimilation*, Frontiers in Applied
Dynamical Systems: Reviews and Tutorials 2, DOI 10.1007/978-3-319-18347-3_2

$$\pi^n(z) = \frac{1}{C} G^n(z) \int_{\mathscr{Z}} \pi(z|z')\pi^{n-1}(z')\mathrm{d}z' \qquad (2.1)$$

with the constant C chosen such that $\int_{\mathscr{Z}} \pi^n(z)\mathrm{d}z = 1$.

A general mathematical framework for such problems is provided by the Feynman-Kac formalism as discussed in detail in del Moral (2004).[1] In order to apply Monte Carlo methods to (2.1) it is useful to reformulate (2.1) in terms of modified Markov processes with transition kernel $\pi^n(z|z')$, which satisfy the consistency condition

$$\pi^n(z) = \int_{\mathscr{Z}} \pi^n(z|z')\pi^{n-1}(z')\mathrm{d}z'. \qquad (2.2)$$

This reformulation has been called the McKean approach to Feynman-Kac models in del Moral (2004).[2] Once a particular McKean model is available, a Monte Carlo implementation reduces to sequences of particles $\{z_i^n\}_{i=1}^M$ being generated sequentially by

$$z_i^n \sim \pi^n(\cdot|z_i^{n-1}), \qquad i = 1, \ldots, M, \qquad (2.3)$$

for $n = 0, 1, \ldots, N$. In other words, z_i^n is the realization of a random variable with (conditional) PDF $\pi^n(z|z_i^{n-1})$. Such a Monte Carlo method constitutes a particular instance of the far more general class of sequential Monte Carlo methods (SMCMs) (Doucet et al., 2001).

While there are many applications that naturally give rise to Feynman-Kac formulae (del Moral, 2004), we will focus in this paper on Markov processes for which the underlying transition kernel $\pi(z|z')$ is determined by a deterministic dynamical system and that we wish to estimate its current state z^n from partial and noisy observations y_{obs}^n. The weight function $G^n(z)$ of a Feynman-Kac recursion (2.1) is in this case given by the likelihood of observing y_{obs}^n given z^n and we encounter a particular application of *Bayesian inference* (Jazwinski, 1970; Stuart, 2010). The precise mathematical setting and the Feynman-Kac formula for the associated data assimilation problem will be discussed in Section 2. Some of the standard Monte Carlo approaches to Feynman-Kac formulae will be summarized in Section 3.

---

[1]The classic Feynman-Kac formulae provide a connection between stochastic processes and solutions to partial differential equations. Here we use a generalization which links discrete-time stochastic processes to sequences of marginal distributions and associated expectation values. In addition to sequential Bayesian inference, which primarily motivates this review article, applications of discrete-time Feynman-Kac formula of type (2.1) can, for example, be found in non-equilibrium molecular dynamics, where the weight functions G^n in (2.1) correspond to the incremental work exerted on a molecular system at time t_n. See Lelièvre et al. (2010) for more details.

[2]McKean (1966) pioneered the study of stochastic processes which are generated by stochastic differential equations for which the diffusion term depends on the time-evolving marginal distributions $\pi(z, t)$. Here we utilize a generalization of this idea to discrete-time Markov processes which allows for transition kernels $\pi^n(z|z')$ to depend on the marginal distributions $\pi^n(z)$.

It is important to note that the consistency condition (2.2) does not specify a McKean model uniquely. In other words, given a Feynman-Kac recursion (2.1) there are many options to define an associated McKean model $\pi^n(z|z')$. It has been suggested independently by Reich (2011); Reich and Cotter (2013) and Moselhy and Marzouk (2012) in the context of Bayesian inference that optimal transportation (Villani, 2003) can be used to couple the prior and posterior distributions. This idea generalizes to all Feynman-Kac formulae and leads to optimal in the sense of optimal transportation McKean models. This optimal transportation approach to McKean models will be developed in detail in Section 4 of this paper.

Optimal transportation problems lead to a nonlinear elliptic PDE, called the Monge-Ampere equation (Villani, 2003), which is very hard to tackle numerically in space dimensions larger than one. On the other hand, optimal transportation is an infinite-dimensional generalization (McCann, 1995; Villani, 2009) of the classic linear transport problem (Strang, 1986). This interpretation is very attractive in terms of Monte Carlo methods and gives rise to a novel SMCM of type (2.3), which we call the ensemble transform particle filter (ETPF) (Reich, 2013a). The ETPF is based on a linear transformation of the forecast particles

$$z_i^f \sim \pi(\cdot|z_i^{n-1}), \qquad i = 1, \ldots, M, \tag{2.4}$$

of type

$$z_j^n = \sum_{i=1}^{M} z_i^f s_{ij} \tag{2.5}$$

with the entries $s_{ij} \geq 0$ of the transform matrix $S \in \mathbb{R}^{M \times M}$ being determined by an appropriate linear transport problem. Even more remarkably, it turns out that SMCMs which resample in each iteration as well as the popular class of ensemble Kalman filters (EnKFs) (Evensen, 2006) also fit into the linear transform framework of (2.5). We will discuss particle/ensemble-based sequential data assimilation algorithms within the unifying framework of linear ensemble transform filters in Section 5. An extension of the ETPF to spatially extended dynamical systems using the concept of localization (Evensen, 2006) is proposed in Section 6. Section 7 provides numerical results for the Lorenz-63 (Lorenz, 1963) and the Lorenz-96 (Lorenz, 1996) models. The results for the 40 dimensional Lorenz-96 indicate that the ensemble transform particle filter with localization can beat the curse of dimensionality which has so far prevented SMCMs from being used for high-dimensional systems (Bengtsson et al., 2008). A brief historical account of data assimilation and filtering is given in Section 7.

We mention that, while the focus of this paper is on state estimation for deterministic dynamical systems, the results can easily be extended to stochastic models as well as combined state and parameter estimation problems. Furthermore, possible applications include all areas in which SMCMs have successfully been used. We refer, for example, to navigation, computer vision, and cognitive sciences (see, e.g., Doucet et al. (2001); Lee and Mumford (2003) and the references therein).

2   Data assimilation and Feynman-Kac formula

Consider a deterministic *dynamical system*[3]

$$z^{n+1} = \Psi(z^n) \tag{2.6}$$

with state variable $z \in \mathbb{R}^{N_z}$, iteration index $n \geq 0$, and given initial $z^0 \in \mathscr{Z} \subset \mathbb{R}^{N_z}$. We assume that Ψ is a diffeomorphism on $\mathbb{R}^{N_z}$ and that $\Psi(\mathscr{Z}) \subseteq \mathscr{Z}$, which implies that the iterates z^n stay in $\mathscr{Z}$ for all $n \geq 0$. Dynamical systems of the form (2.6) often arise as the time-Δt-flow maps of differential equations

$$\frac{dz}{dt} = f(z). \tag{2.7}$$

In many practical applications the initial state z^0 is not precisely known. We may then assume that our uncertainty about the correct initial state can, for example, be quantified in terms of ratios of frequencies of occurrence. More precisely, the *ratio of frequency of occurrence* (RFO) of two initial conditions $z_a^0 \in \mathscr{Z}$ and $z_b^0 \in \mathscr{Z}$ is defined as the RFOs for the two associated ε-neighborhoods $\mathscr{U}_\varepsilon(z_a^0)$ and $\mathscr{U}_\varepsilon(z_b^0)$, respectively, and taking the limit $\varepsilon \to 0$. Here $\mathscr{U}_\varepsilon(z)$ is defined as

$$\mathscr{U}_\varepsilon(z) = \{z' \in \mathbb{R}^{N_z} : \|z' - z\|^2 \leq \varepsilon\}$$

It is important to note that the volume of both neighborhoods is identical, *i.e.*, $V(\mathscr{U}_\varepsilon(z_a^0)) = V(\mathscr{U}_\varepsilon(z_b^0))$.

From a frequentist perspective, RFOs can be thought of as arising from repeated experiments with one and the same dynamical system (2.6) under varying initial conditions and upon counting how often $z^0 \in \mathscr{U}_\varepsilon(z_a^0)$ relative to $z^0 \in \mathscr{U}_\varepsilon(z_b^0)$. There will, of course, be many instances for which z^0 is neither in $\mathscr{U}_\varepsilon(z_a^0)$ nor $\mathscr{U}_\varepsilon(z_b^0)$. Alternatively, one can take a Bayesian perspective and think of RFOs as our subjective belief about a z_a^0 to actually arise as the initial condition in (2.6) relative to another initial condition z_b^0. The later interpretation is also applicable in case only a single experiment with (2.6) is conducted.

Independent of such a statistical interpretation of the RFO, we assume the availability of a function $\tau(z) > 0$ such that the RFO can be expressed as

$$\text{RFO} = \frac{\tau(z_a^0)}{\tau(z_b^0)} \tag{2.8}$$

for all pairs of initial conditions from $\mathscr{Z}$.

---

[3]Even though this review article assumes deterministic evolution equations, the results presented here can easily be generalized to evolution equations with stochastic model errors.

Provided that

$$\int_{\mathscr{Z}} \tau(z)\mathrm{d}z < \infty$$

we can introduce the probability density function (PDF)

$$\pi_{Z^0}(z) = \frac{\tau(z)}{\int_{\mathscr{Z}} \tau(z)\mathrm{d}z}$$

and interpret initial conditions z^0 as realizations of a random variable $Z^0 : \Omega \to \mathbb{R}^{N_z}$ with PDF π_{Z^0}.[4] We remark that most of our subsequent discussions carry through even if $\int_{\mathscr{Z}} \tau(z)\mathrm{d}z$ is unbounded as long as the RFOs remain well-defined.

So far we have discussed RFOs for initial conditions. But one can also consider such ratios for any iteration index $n \geq 0$, *i.e.*, for solutions

$$z_a^n = \Psi^n(z_a^0)$$

and

$$z_b^n = \Psi^n(z_b^0).$$

Here Ψ^n denotes the n-fold application of Ψ. The RFO at iteration index n is now defined as the ratio of the frequencies of occurrence for the two associated ε-neighborhoods $\mathscr{U}_\varepsilon(z_a^n)$ and $\mathscr{U}_\varepsilon(z_b^n)$, respectively, in the limit $\varepsilon \to 0$. We pull this ratio back to $n = 0$ and find that

$$\mathrm{RFO}(n) \approx \frac{\tau(\Psi^{-n}(z_a^n))V_a}{\tau(\Psi^{-n}(z_b^n))V_b}$$

for ε sufficiently small, where

$$V_{a/b} = V(\Psi^{-n}(\mathscr{U}_\varepsilon(z_{a/b}^n))) := \int_{\Psi^{-n}(\mathscr{U}_\varepsilon(z_{a/b}^n))} \mathrm{d}z$$

denote the volumes of $\Psi^{-n}(\mathscr{U}_\varepsilon(z_{a/b}^n))$ and Ψ^{-n} refers to the inverse of Ψ^n. These two volumes can be approximated as

$$V_{a/b} \approx V(\mathscr{U}_\varepsilon(z_{a/b}^0)) \times |D\Psi^{-n}(z_{a/b}^n)|$$

----

[4] We have assumed the existence of an underlying probability space $(\Omega, \mathscr{F}, \mathbb{P})$. The specific structure of this probability space does not play a role in the subsequent discussions.

for $\varepsilon > 0$ sufficiently small. Here $D\Psi^{-n}(z) \in \mathbb{R}^{N_z \times N_z}$ stands for the Jacobian matrix of partial derivatives of Ψ^{-n} at z and $|D\Psi^{-n}(z)|$ for its determinant. Hence, upon taking the limit $\varepsilon \to 0$, we obtain

$$
\begin{aligned}
\text{RFO}(n) &= \frac{\tau(\Psi^{-n}(z_a^n))|D\Psi^{-n}(z_a^n)|}{\tau(\Psi^{-n}(z_b^n))|D\Psi^{-n}(z_b^n)|} \\
&= \frac{\pi_{Z^0}(\Psi^{-n}(z_a^n))|D\Psi^{-n}(z_a^n)|}{\pi_{Z^0}(\Psi^{-n}(z_b^n))|D\Psi^{-n}(z_b^n)|}.
\end{aligned}
$$

Therefore we may interpret solutions z^n for fixed iteration index $n \geq 1$ as realizations of a random variable $Z^n : \Omega \to \mathbb{R}^{N_z}$ with PDF

$$
\pi_{Z^n}(z) = \pi_{Z^0}(\Psi^{-n}(z))|D\Psi^{-n}(z)|. \tag{2.9}
$$

These PDFs can also be defined recursively using

$$
\pi_{Z^{n+1}}(z) = \int_{\mathscr{Z}} \delta(z - \Psi(z')) \, \pi_{Z^n}(z') \, \mathrm{d}z'. \tag{2.10}
$$

Here $\delta(\cdot)$ denotes the Dirac delta function, which satisfies

$$
\int_{\mathbb{R}^{N_z}} f(z)\delta(z - \bar{z})\mathrm{d}z = f(\bar{z})
$$

for all smooth functions $f : \mathbb{R}^{N_z} \to \mathbb{R}$.[5] In other words, the dynamical system (2.6) induces a *Markov process*, which we can also write as

$$
Z^{n+1} = \Psi(Z^n)
$$

in terms of the random variables Z^n, $n \geq 0$.

The sequence of random variables $\{Z^n\}_{n=0}^N$ for fixed $N \geq 1$ gives rise to the finite-time stochastic process $Z^{0:N} : \Omega \to \mathscr{Z}^{N+1}$ with realizations

$$
z^{0:N} := (z^0, z^1, \dots, z^N) = Z^{0:N}(\omega), \quad \omega \in \Omega,
$$

that satisfy (2.6). The joint distribution of $Z^{0:N}$, denoted by $\pi_{Z^{0:N}}$, is formally[6] given by

$$
\pi_{Z^{0:N}}(z^0, \dots, z^N) = \pi_{Z^0}(z^0) \, \delta(z^1 - \Psi(z^0)) \cdots \delta(z^N - \Psi(z^{N-1})) \tag{2.11}
$$

and (2.9) is the marginal of $\pi_{Z^{0:N}}$ in z^n, $n = 1, \dots, N$.

---

[5]The Dirac delta function $\delta(z - \bar{z})$ provides a convenient shorthand for the point measure $\mu_{\bar{z}}(\mathrm{d}z)$.

[6]To be mathematically precise one should talk about the joint measure

$$
\mu_{Z^{0:N}}(\mathrm{d}z^0, \dots, \mathrm{d}z^N) = \mu_{Z^0}(\mathrm{d}z^0)\mu_{\Psi(z^0)}(\mathrm{d}z^1) \cdots \mu_{\Psi(z^{N-1})}(\mathrm{d}z^N)
$$

with initial measure $\mu_{Z^0}(\mathrm{d}z^0) = \pi_{Z^0}(z^0)\mathrm{d}z^0$.

Let us now consider the situation where (2.6) serves as a model for an unknown physical process with realization

$$z_{\text{ref}}^{0:N} = (z_{\text{ref}}^0, z_{\text{ref}}^1, \ldots, z_{\text{ref}}^N). \qquad (2.12)$$

In the classic filtering/smoothing setting (Jazwinski, 1970; Bain and Crisan, 2009) one assumes that there exists an $\omega_{\text{ref}} \in \Omega$ such that

$$z_{\text{ref}}^{0:N} = Z^{0:N}(\omega_{\text{ref}}).$$

In practice such an assumption is highly unrealistic and the reference trajectory (2.12) may instead follow an iteration

$$z_{\text{ref}}^{n+1} = \Psi_{\text{ref}}(z_{\text{ref}}^n) \qquad (2.13)$$

with unknown initial z_{ref}^0 and unknown Ψ_{ref}. Of course, it should hold that Ψ in (2.6) is close to Ψ_{ref} in an appropriate mathematical sense.

Independently of such assumptions, we assume that $z_{\text{ref}}^{0:N}$ is accessible to us through partial and noisy observations of the form

$$y_{\text{obs}}^n = h(z_{\text{ref}}^n) + \xi^n, \quad n = 1, \ldots, N,$$

where $h : \mathscr{Z} \to \mathbb{R}^{N_y}$ is called the *forward* or observation map and the ξ^n's are realizations of independent and identically distributed Gaussian random variables with mean zero and covariance matrix $R \in \mathbb{R}^{N_y \times N_y}$. Estimating z_{ref}^n from y_{obs}^n constitutes a classic inverse problem (Tarantola, 2005).

The *ratio of fits to data* (RFD) of two realizations $z_a^{0:N}$ and $z_b^{0:N}$ from the stochastic process $Z^{0:N}$ is defined as

$$\text{RFD} = \frac{\prod_{n=1}^N e^{-\frac{1}{2}(h(z_a^n)-y_{\text{obs}}^n)^T R^{-1}(h(z_a^n)-y_{\text{obs}}^n)}}{\prod_{n=1}^N e^{-\frac{1}{2}(h(z_b^n)-y_{\text{obs}}^n)^T R^{-1}(h(z_b^n)-y_{\text{obs}}^n)}}.$$

Finally we define the *ratio of fits to model and data* (RFMD) of a $z_a^{0:N}$ versus a $z_b^{0:N}$ given the model and the observations as follows:

$$\begin{aligned}
\text{RFMD} &= \text{RFD} \times \text{RFO}(0) \\
&= \frac{\prod_{n=1}^N e^{-\frac{1}{2}(h(z_a^n)-y_{\text{obs}}^n)^T R^{-1}(h(z_a^n)-y_{\text{obs}}^n)}}{\prod_{n=1}^N e^{-\frac{1}{2}(h(z_b^n)-y_{\text{obs}}^n)^T R^{-1}(h(z_b^n)-y_{\text{obs}}^n)}} \frac{\pi_{Z^0}(z_a^0)}{\pi_{Z^0}(z_b^0)} \\
&= \frac{\prod_{n=1}^N e^{-\frac{1}{2}(h(z_a^n)-y_{\text{obs}}^n)^T R^{-1}(h(z_a^n)-y_{\text{obs}}^n)}}{\prod_{n=1}^N e^{-\frac{1}{2}(h(z_b^n)-y_{\text{obs}}^n)^T R^{-1}(h(z_b^n)-y_{\text{obs}}^n)}} \frac{\pi_{Z^{0:N}}(z_a^{0:N})}{\pi_{Z^{0:N}}(z_b^{0:N})}. \qquad (2.14)
\end{aligned}$$

The simple product structure arises since the uncertainty in the initial conditions is assumed to be independent of the measurement errors and the last identity follows from the fact that our dynamical model is deterministic.

Again we may translate this combined ratio into a PDF

$$\pi_{Z^{0:N}}(z^{0:N}|y_{\text{obs}}^{1:N}) = \frac{1}{C}\prod_{n=1}^{N}e^{-\frac{1}{2}(h(z^n)-y_{\text{obs}}^n)^T R^{-1}(h(z^n)-y_{\text{obs}}^n)}\pi_{Z^0}(z^0), \qquad (2.15)$$

where $C > 0$ is a normalization constant depending only on $y_{\text{obs}}^{1:N}$. This PDF gives the probability distribution in $z^{0:N}$ conditioned on the given set of observations

$$y_{\text{obs}}^{1:N} = (y_{\text{obs}}^1, \ldots, y_{\text{obs}}^N).$$

The PDF (2.15) is, of course, also conditioned on (2.6) and the initial PDF π_{Z^0}. This dependence is not explicitly taken account of in order to avoid additional notational clutter.

The formulation (2.15) is an instance of Bayesian inference on the one hand, and an instance of the Feynman-Kac formalism on the other. Within the Bayesian perspective, π_{Z^0} (or, equivalently, $\pi_{Z^{0:N}}$) represents the prior distribution,

$$\pi_{Y^{1:N}}(y^{1:n}|z^{0:n}) = \frac{1}{(2\pi)^{N_y N/2}|R|^{N/2}}\prod_{n=1}^{N}e^{-\frac{1}{2}(h(z^n)-y^n)^T R^{-1}(h(z^n)-y^n)}$$

the compounded likelihood function, and (2.15) the posterior PDF given an actually observed $y^{1:n} = y_{\text{obs}}^{1:n}$. The Feynman-Kac formalism is more general and includes a wide range of applications for which an underlying stochastic process is modified by weights $G^n(z^n) \geq 0$. These weights then replace the likelihood functions

$$\pi_Y(y_{\text{obs}}^n|z^n) = \frac{1}{(2\pi)^{N_y/2}|R|^{1/2}}e^{-\frac{1}{2}(h(z^n)-y_{\text{obs}}^n)^T R^{-1}(h(z^n)-y_{\text{obs}}^n)}$$

in (2.15). The functions $G^n : \mathbb{R}^{N_z} \to \mathbb{R}$ can depend on the iteration index, as in

$$G^n(z) := \pi_Y(y_{\text{obs}}^n|z)$$

or may be independent of the iteration index. See del Moral (2004) for further details on the Feynman-Kac formalism and Lelièvre et al. (2010) for a specific (non-Bayesian) application in the context of non-equilibrium molecular dynamics.

Formula (2.15) is hardly ever explicitly accessible and one needs to resort to numerical approximations whenever one wishes to either compute the expectation value

$$\mathbb{E}[f(Z^{0:N})|y_{\text{obs}}^{1:N}] = \int_{\mathscr{Z}^{N+1}}f(z^{0:N})\,\pi_{Z^{0:N}}(z^{0:N}|y_{\text{obs}}^{1:N})\mathrm{d}z^0\cdots\mathrm{d}z^N$$

of a given function $f : \mathscr{Z}^{N+1} \to \mathbb{R}$ or the

$$\text{RFMD} = \frac{\pi_{Z^{0:N}}(z_a^{0:N}|y_{\text{obs}}^{1:N})}{\pi_{Z^{0:N}}(z_b^{0:N}|y_{\text{obs}}^{1:N})}$$

for given trajectories $z_a^{0:N}$ and $z_b^{0:N}$. Basic Monte Carlo approximation methods will be discussed in Section 3. Alternatively, one may seek the *maximum a posteriori (MAP) estimator* z_{MAP}^0, which is defined as the initial condition z^0 that maximizes (2.15) or, formulated alternatively,

$$z_{MAP}^0 = \arg\inf L(z^0), \qquad L(z^0) := -\log \pi_{Z^{0:N}}(z^{0:N}|y_{obs}^{1:N})$$

(Kaipio and Somersalo, 2005; Lewis et al., 2006; Tarantola, 2005). The MAP estimator is closely related to *variational data assimilation techniques*, such as 3D-Var and 4D-Var, widely used in meteorology (Daley, 1993; Kalnay, 2002).

In many applications expectation values need to be computed for functions f which depend only on a single z^n. Those expectation values can be obtained by first integrating out all components in (2.15) except for z^n. We denote the resulting marginal PDF by $\pi_{Z^n}(z^n|y_{obs}^{1:N})$. The case $n = 0$ plays a particular role since

$$\text{RFMD} = \frac{\pi_{Z^0}(z_a^0|y_{obs}^{1:N})}{\pi_{Z^0}(z_b^0|y_{obs}^{1:N})}$$

and

$$\pi_{Z^n}(z^n|y_{obs}^{1:N}) = \pi_{Z^0}(\Psi^{-n}(z^n)|y_{obs}^{1:N})|D\Psi^{-n}(z^n)|$$

for $n = 1, \ldots, N$. These identities hold because our dynamical system (2.6) is deterministic and invertible. In Section 4 we will discuss recursive approaches for determining the marginal $\pi_{Z^N}(z^N|y_{obs}^{1:N})$ in terms of Markov processes. Computational techniques for implementing such recursions will be discussed in Section 5.

3 Monte Carlo methods in path space

In this section, we briefly summarize two popular Monte Carlo strategies for computing expectation values with respect to the complete conditional distribution $\pi_{Z^{0:N}}(\cdot|y_{obs}^{1:N})$. We start with the classic importance sampling Monte Carlo method.

3.1 Ensemble prediction and importance sampling

Ensemble prediction is a Monte Carlo method for assessing the marginal PDFs (2.9) for $n = 0, \ldots, N$. One first generates z_i^0, $i = 1, \ldots, M$, independent samples from the initial PDF π_{Z^0}; *i.e.*, samples are generated such that the probability of being in $\mathcal{U}_\varepsilon(z)$ is

$$\int_{\mathcal{U}_\varepsilon(z)} \pi_{Z^0}(z')dz' \approx V(\mathcal{U}_\varepsilon(z)) \times \pi_{Z^0}(z).$$

Furthermore, the expectation value of a function f with respect to Z^0 is approximated by the familiar empirical estimator

$$\bar{f}_M := \frac{1}{M} \sum_{i=1}^{M} f(z_i^0).$$

The initial ensemble $\{z_i^0\}_{i=1}^{M}$ is propagated independently under the dynamical system (2.6) for $n = 0, \ldots, N - 1$. This yields M trajectories

$$z_i^{0:N} = (z_i^0, z_i^1, \ldots, z_i^N),$$

which provide independent samples from the $\pi_{Z^{0:N}}$ distribution. With each of these samples we associate the weight

$$w_i = \frac{1}{C} \prod_{n=1}^{N} e^{-\frac{1}{2}(h(z_i^n) - y_{\text{obs}}^n)^T R^{-1} (h(z_i^n) - y_{\text{obs}}^n)}$$

with the constant of proportionality chosen such that $\sum_{i=1}^{M} w_i = 1$.

The RFMD for any pair of samples $z_i^{1:N}$ and $z_j^{1:N}$ from $\pi_{Z^{0:N}}$ is now simply given by w_i / w_j and the expectation value of a function f with respect to $\pi_{Z^{0:N}}(z^{0:N}|y_{\text{obs}}^{1:N})$ can be approximated by the empirical estimator

$$\bar{f}_M = \sum_{i=1}^{M} w_i f(z_i^{0:N}).$$

This estimator is an instance of *importance sampling* since samples from a distribution different from the target distribution, here $\pi_{Z^{0:N}}$, are used to approximate the statistics of the target distribution, here $\pi_{Z^{0:N}}(\cdot|y_{\text{obs}}^{1:N})$. See Liu (2001) and Robert and Casella (2004) for further details.

3.2   Markov chain Monte Carlo (MCMC) methods

Importance sampling becomes inefficient whenever the *effective sample size*

$$M_{\text{eff}} = \frac{1}{\sum_{i=1}^{M} w_i^2} \in [1, M] \qquad (2.16)$$

becomes much smaller than the sample size M. Under those circumstances it can be preferable to generate dependent samples $z_i^{0:N}$ using MCMC methods. MCMC methods rely on a proposal step and a Metropolis-Hastings acceptance criterion.

Note that only z_i^0 needs to be stored since the whole trajectory is then uniquely determined by $z_i^n = \Psi^n(z_i^0)$. Consider, for simplicity, the reversible proposal step

$$z_p^0 = z_i^0 + \xi,$$

where ξ is a realization of a random variable Ξ with PDF π_Ξ satisfying $\pi_\Xi(\xi) = \pi_\Xi(-\xi)$ and z_i^0 denotes the last accepted sample. The associated trajectory $z_p^{0:N}$ is computed using (2.6). Next one evaluates the RFMD (2.14) for $z_a^{0:N} = z_p^{0:N}$ versus $z_b^{0:N} = z_i^{0:N}$, i.e.,

$$\alpha := \frac{\prod_{n=1}^N e^{-\frac{1}{2}(h(z_p^n)-y_{obs}^n)^T R^{-1}(h(z_p^n)-y_{obs}^n)}\,\pi_{Z^0}(z_p^0)}{\prod_{n=1}^N e^{-\frac{1}{2}(h(z_i^n)-y_{obs}^n)^T R^{-1}(h(z_i^n)-y_{obs}^n)}\,\pi_{Z^0}(z_i^0)}.$$

If $\alpha \geq 1$, then the proposal is accepted and the new sample for the initial condition is $z_{i+1}^0 = z_p^0$. Otherwise the proposal is accepted with probability α and rejected with probability $1 - \alpha$. In case of rejection one sets $z_{i+1}^0 = z_i^0$. Note that the accepted samples follow the $\pi_{Z^0}(z^0|y_{obs}^{1:N})$ distribution and not the initial PDF $\pi_{Z^0}(z^0)$.

A potential problem with MCMC methods lies in low acceptance rates and/or highly dependent samples. In particular, if the distribution in π_Ξ is too narrow, then exploration of phase space can be slow while a too wide distribution can potentially lead to low acceptance rates. Hence one should compare the effective sample size (2.16) from an importance sampling approach to the effective sample size of an MCMC implementation, which is inversely proportional to the integrated autocorrelation time of the samples. See Liu (2001) and Robert and Casella (2004) for more details.

We close this section by referring to Särkkä (2013) for further algorithms for filtering and smoothing.

4 McKean optimal transportation approach

We now restrict the discussion of the Feynman-Kac formalism to the marginal PDFs $\pi_{Z^n}(z^n|y_{obs}^{1:N})$. We have already seen in Section 2 that the marginal PDF with $n = N$ plays a particularly important role. We show that this marginal PDF can be recursively defined starting from the PDF π_{Z^0} for the initial condition z^0 of (2.6). For that reason we introduce the forecast and analysis PDF at iteration index n and denote them by $\pi_{Z^n}(z^n|y_{obs}^{1:n-1})$ and $\pi_{Z^n}(z^n|y_{obs}^{1:n})$, respectively. Those PDFs are defined recursively by

$$\pi_{Z^n}(z^n|y_{obs}^{1:n-1}) = \pi_{Z^{n-1}}(\Psi^{-1}(z^n)|y_{obs}^{1:n-1})|D\Psi^{-1}(z^n)| \tag{2.17}$$

and

$$\pi_{Z^n}(z^n|y_{obs}^{1:n}) = \frac{\pi_Y(y_{obs}^n|z^n)\pi_{Z^n}(z^n|y_{obs}^{1:n-1})}{\int_{\mathscr{Z}} \pi_Y(y_{obs}^n|z)\pi_{Z^n}(z|y_{obs}^{1:n-1})dz}. \tag{2.18}$$

Here (2.17) simply propagates the analysis from index $n - 1$ to the forecast at index n under the action of the dynamical system (2.6). Bayes' formula is then applied in (2.18) in order to transform the forecast into the analysis at index n by assimilating the observed y_{obs}^n.

Theorem 1. *Consider the sequence of forecast and analysis PDFs defined by the recursion (2.17)–(2.18) for $n = 1, \ldots, N$ with $\pi_{Z^0}(z^0)$ given. Then the analysis PDF at $n = N$ is equal to the Feynman-Kac PDF (2.15) marginalized down to the single variable z^N.*

Proof. We prove the theorem by induction in N. We first verify the claim for $N = 1$. Indeed, by definition,

$$
\begin{aligned}
\pi_{Z^1}(z^1|y_{obs}^1) &= \frac{1}{C_1} \int_{\mathscr{L}} \pi_Y(y_{obs}^1|z^1)\delta(z^1 - \Psi(z^0))\pi_{Z^0}(z^0)dz^0 \\
&= \frac{1}{C_1}\pi_Y(y_{obs}^1|z^1)\pi_{Z^0}(\Psi^{[-1]}(z^1))|D\Psi^{-1}(z^1)| \\
&= \frac{1}{C_1}\pi_Y(y_{obs}^1|z^1)\pi_{Z^1}(z^1)
\end{aligned}
$$

and $\pi_{Z^1}(z^1)$ is the forecast PDF at index $n = 1$. Here C_1 denotes the constant of proportionality which only depends on y_{obs}^1.

The induction step from N to $N + 1$ follows from the following line of arguments. We know by the induction assumption that the marginal PDF at $N + 1$ can be written as

$$
\begin{aligned}
\pi_{Z^{N+1}}(z^{N+1}|y_{obs}^{1:N+1}) &= \int_{\mathscr{L}} \pi_{Z^{N:N+1}}(z^{N:N+1}|y_{obs}^{1:N+1})dz^N \\
&= \frac{1}{C_{N+1}}\pi_Y(y_{obs}^{N+1}|z^{N+1}) \int_{\mathscr{L}} \delta(z^{N+1} - \Psi(z^N))\pi_{Z^N}(z^N|y_{obs}^{1:N})dz^N \\
&= \frac{1}{C_{N+1}}\pi_Y(y_{obs}^{N+1}|z^{N+1})\pi_{Z^{N+1}}(z^{N+1}|y_{obs}^{1:N})
\end{aligned}
$$

in agreement with (2.18) for $n = N + 1$. Here C_{N+1} denotes the constant of proportionality that depends only on y_{obs}^{N+1} and we have made use of the fact that the forecast PDF at index $n = N + 1$ is, according to (2.17), defined by

$$
\pi_{Z^{N+1}}(z^{N+1}|y_{obs}^{1:N}) = \pi_{Z^N}(\Psi^{-1}(z^{N+1})|y_{obs}^{1:N})|D\Psi^{-1}(z^{N+1})|.
$$

$\square$

While the forecast step (2.17) is in the form of a Markov process with transition kernel

$$
\pi_{model}(z|z') = \delta(z - \Psi(z')),
$$

this does not hold for the analysis step (2.18). The *McKean approach* to (2.17)–(2.18) is based on introducing Markov transition kernels $\pi_{\text{data}}^n(z|z')$, $n = 1, \ldots, N$, for the analysis step (2.18). In other words, the transition kernel π_{data}^n at iteration index n has to satisfy the consistency condition

$$\pi_{Z^n}(z|y_{\text{obs}}^{1:n}) = \int_{\mathscr{L}} \pi_{\text{data}}^n(z|z')\pi_{Z^n}(z'|y_{\text{obs}}^{1:n-1})\mathrm{d}z'. \tag{2.19}$$

These transition kernels are not unique. The following kernel

$$\pi_{\text{data}}^n(z|z') = \epsilon\pi_Y(y_{\text{obs}}^n|z')\delta(z-z') + \left(1 - \epsilon\pi_Y(y_{\text{obs}}^n|z')\right)\pi_{Z^n}(z|y_{\text{obs}}^{1:n}) \tag{2.20}$$

has, for example, been considered in del Moral (2004). Here $\epsilon \geq 0$ has to be chosen such that

$$1 - \epsilon\pi_Y(y_{\text{obs}}^n|z) \geq 0$$

for all $z \in \mathbb{R}^{N_z}$. Indeed, we find that

$$\int_{\mathscr{L}} \pi_{\text{data}}(z|z')\pi_{Z^n}(z'|y_{\text{obs}}^{1:n-1})\mathrm{d}z' = \pi_{Z^n}(z|y_{\text{obs}}^{1:n}) + \epsilon\pi_Y(y_{\text{obs}}^n|z)\pi_{Z^n}(z|y_{\text{obs}}^{1:n-1}) -$$

$$\epsilon\pi_{Z^n}(z|y_{\text{obs}}^{1:n})\pi_Y(y_{\text{obs}}^n|y_{\text{obs}}^{1:n-1})$$

$$= \pi_{Z^n}(z|y_{\text{obs}}^{1:n}).$$

The intuitive interpretation of this transition kernel is that one stays at z' with probability $p = \epsilon\pi_Y(y_{\text{obs}}^n|z')$ while with probability $(1 - p)$ one samples from the analysis PDF $\pi_{Z^n}(z|y_{\text{obs}}^{1:n})$.

Let us define the combined McKean-Markov transition kernel

$$\pi^n(z^n|z^{n-1}) := \int_{\mathscr{L}} \pi_{\text{data}}^n(z^n|z)\pi_{\text{model}}(z|z^{n-1})\mathrm{d}z$$

$$= \int_{\mathscr{L}} \pi_{\text{data}}^n(z^n|z)\delta(z - \Psi(z^{n-1}))\mathrm{d}z$$

$$= \pi_{\text{data}}^n(z^n|\Psi(z^{n-1})) \tag{2.21}$$

for the propagation of the analysis PDF from iteration index $n-1$ to n. The combined McKean-Markov transition kernels π^n, $n = 1, \ldots, N$, define a finite-time stochastic process $\hat{Z}^{0:N} = \{\hat{Z}^n\}_{n=0}^N$ with $\hat{Z}^0 = Z^0$. The marginal PDFs satisfy

$$\pi_{\hat{Z}^n}(z^n|y_{\text{obs}}^{1:n}) = \int_{\mathscr{L}} \pi^n(z^n|z^{n-1})\pi_{\hat{Z}^{n-1}}(z^{n-1}|y_{\text{obs}}^{1:n-1})\mathrm{d}z^{n-1}$$

$$= \pi_{Z^n}(z^n|y_{\text{obs}}^{1:n})$$

Corollary 1. *The final time marginal distribution* $\pi_{Z^N}(z^N|y_{\mathrm{obs}}^{1:N})$ *of the Feynman-Kac formulation (2.15) is identical to the final time marginal distribution* $\pi_{\hat{Z}^N}(z^N|y_{\mathrm{obs}}^{1:N})$ *of the finite-time stochastic process* $\hat{Z}^{0:N}$ *induced by the McKean-Markov transition kernels* $\pi^n(z^n|z^{n-1})$.

We summarize our discussion on the McKean approach in terms of analysis and forecast random variables, which constitute the basic building blocks for most current sequential data assimilation methods.

Definition 1. Given a dynamic iteration (2.6) with PDF π_{Z^0} for the initial conditions and observations y_{obs}^n, $n = 1, \ldots, N$, the McKean approach leads to a recursive definition of forecast $Z^{n,f}$ and analysis $Z^{n,a}$ random variables. The iteration is started by declaring Z^0 the analysis $Z^{0,a}$ at $n = 0$. The forecast at iteration index $n > 0$ is defined by propagating the analysis at $n - 1$ forward under the dynamic model (2.6), *i.e.*,

$$Z^{n,f} = \Psi(Z^{n-1,a}). \qquad (2.22)$$

The analysis $Z^{n,a}$ at iteration index n is defined by applying the transition kernel $\pi_{\mathrm{data}}^n(z|z')$ to $Z^{n,f}$. In particular, if $z^{n,f} = Z^{n,f}(\omega)$ is a realized forecast at iteration index n, then the analysis is distributed according to

$$Z^{n,a}|z^{n,f} \sim \pi_{\mathrm{data}}^n(z|z^{n,f}). \qquad (2.23)$$

If the marginal PDFs of $Z^{n,f}$ and $Z^{n,a}$ are denoted by $\pi_{Z^{n,f}}(z)$ and $\pi_{Z^{n,a}}(z)$, respectively, then the transition kernel π_{data}^n has to satisfy the compatibility condition (2.19), *i.e.*,

$$\int_{\mathscr{Z}} \pi_{\mathrm{data}}^n(z|z')\pi_{Z^{n,f}}(z')\mathrm{d}z' = \pi_{Z^{n,a}}(z) \qquad (2.24)$$

with

$$\pi_{Z^{n,a}}(z) = \frac{\pi_Y(y_{\mathrm{obs}}^n|z)\pi_{Z^{n,f}}(z)}{\int_{\mathscr{Z}} \pi_Y(y_{\mathrm{obs}}^n|z)\pi_{Z^{n,f}}(z)\mathrm{d}z}.$$

The data related transition step (2.23) introduces randomness into the analysis of a given forecast value. This appears counterintuitive and, indeed, the main purpose of the rest of this section is to demonstrate that the transition kernel π_{data}^n can be chosen such that

$$Z^{n,a} = \nabla_z \phi^n(Z^{n,f}), \qquad (2.25)$$

where $\phi^n : \mathbb{R}^{N_z} \to \mathbb{R}$ is an appropriate convex potential. In other words, the data-driven McKean update step can be reduced to a (deterministic) map and the stochastic process $\hat{Z}^{0:N}$ is induced by the deterministic recursion (or dynamical system)

$$\hat{Z}^n = \nabla_z \phi^n(\Psi(\hat{Z}^{n-1}))$$

with $\hat{Z}^0 = Z^0$.

The compatibility condition (2.24) with $\pi_{\text{data}}^n(z|z') = \delta(z - \nabla_z \phi^n(z'))$ reduces to

$$\pi_{Z^{n,a}}(\nabla_z \phi^n(z))|D\nabla_z \phi^n(z)| = \pi_{Z^{n,f}}(z), \qquad (2.26)$$

which constitutes a highly nonlinear elliptic PDE for the potential ϕ^n. In the remainder of this section we discuss under which conditions this PDE has a solution. This discussion will also guide us towards a numerical approximation technique that circumvents the need for directly solving (2.26) either analytically or numerically.

Consider the forecast PDF $\pi_{Z^{n,f}}$ and the analysis PDF $\pi_{Z^{n,a}}$ at iteration index n. For simplicity of notion we drop the iteration index and simply write π_{Z^f} and π_{Z^a}, respectively.

Definition 2. A *coupling* of π_{Z^f} and π_{Z^a} consists of a pair $Z^{f:a} = (Z^f, Z^a)$ of random variables such that $Z^f \sim \pi_{Z^f}$, $Z^a \sim \pi_{Z^a}$, and $Z^{f:a} \sim \pi_{Z^{f:a}}$. The joint PDF $\pi_{Z^{f:a}}(z^f, z^a)$ on the product space $\mathbb{R}^{N_z} \times \mathbb{R}^{N_z}$ is called the *transference plan* for this coupling. The set of all transference plans is denoted by $\Pi(\pi_{Z^f}, \pi_{Z^a})$.[7]

Clearly, couplings always exist since one can use the trivial product coupling

$$\pi_{Z^{f:a}}(z^f, z^a) = \pi_{Z^f}(z^f)\pi_{Z^a}(z^a),$$

in which case the associated random variables Z^f and Z^a are independent and each random variable follows its given marginal distribution. Once a coupling has been found, a McKean transition kernel is determined by

$$\pi_{\text{data}}(z|z') = \frac{\pi_{Z^{f:a}}(z', z)}{\int_{\mathcal{Z}} \pi_{Z^{f:a}}(z', z'')\mathrm{d}z''}.$$

Reversely, any transition kernel $\pi_{\text{data}}(z|z')$, such as (2.20), also induces a coupling.

A diffeomorphism $T : \mathcal{Z} \to \mathcal{Z}$ is called a *transport map* if the induced random variable $Z^a = T(Z^f)$ satisfies

$$\int_{\mathcal{Z}} f(z^a)\pi_{Z^a}(z^a)\mathrm{d}z^a = \int_{\mathcal{Z}} f(T(z^f))\pi_{Z^f}(z^f)\mathrm{d}z^f$$

for all suitable functions $f : \mathcal{Z} \to \mathbb{R}$. The associated coupling

$$\pi_{Z^{f:a}}(z^f, z^a) = \delta(z^a - T(z^f))\pi_{Z^f}(z^f)$$

is called a *deterministic coupling*. Indeed, one finds that

$$\int_{\mathcal{Z}} \pi_{Z^{f:a}}(z^f, z^a)\mathrm{d}z^a = \pi_{Z^f}(z^f)$$

---

[7]Couplings should be properly defined in terms of probability measures. A coupling between two measures $\mu_{Z^f}(\mathrm{d}z^f)$ and $\mu_{Z^a}(\mathrm{d}z^a)$ consists of a pair of random variables with joint measure $\mu_{Z^{f:a}}(\mathrm{d}z^f, \mathrm{d}z^a)$ such that $\mu_{Z^f}(\mathrm{d}z^f) = \int_{\mathcal{Z}} \mu_{Z^{f:a}}(\mathrm{d}z^f, \mathrm{d}z^a)$ and $\mu_{Z^a}(\mathrm{d}z^a) = \int_{\mathcal{Z}} \mu_{Z^{f:a}}(\mathrm{d}z^f, \mathrm{d}z^a)$, respectively.

and

$$\pi_{Z^a}(z^a) = \int_{\mathscr{Z}} \pi_{Z^{f:a}}(z^f, z^a) \mathrm{d}z^f = \pi_{Z^f}(T^{-1}(z^a))|DT^{-1}(z^a)|,$$

respectively.

When it comes to actually choosing a particular coupling from the set $\Pi(\pi_{Z^f}, \pi_{Z^a})$ of all admissible ones, it appears preferable to pick the one that maximizes the covariance or correlation between Z^f and Z^a. But maximizing their covariance for given marginals has an important geometric interpretation. Consider, for simplicity, univariate random variables Z^f and Z^a, then

$$\mathbb{E}[|Z^f - Z^a|^2] = \mathbb{E}[|Z^f|^2] + \mathbb{E}[|Z^a|^2] - 2\mathbb{E}[Z^f Z^a]$$

$$= \mathbb{E}[|Z^f|^2] + \mathbb{E}[|Z^a|^2] - 2\mathbb{E}[(Z^f - \bar{z}^f)(Z^a - \bar{z}^a)] - 2\bar{z}^f \bar{z}^a$$

$$= \mathbb{E}[|Z^f|^2] + \mathbb{E}[|Z^a|^2] - 2\bar{z}^f \bar{z}^a - 2\mathrm{cov}(Z^f, Z^a),$$

where $\bar{z}^{f/a} = \mathbb{E}[Z^{f/a}]$. Hence finding a joint PDF $\pi_{Z^{f:a}} \in \Pi(\pi_{Z^f}, \pi_{Z^a})$ that minimizes the expectation of $|z^f - z^a|^2$ simultaneously maximizes the covariance between Z^f and Z^a. This geometric interpretation leads to the celebrated Monge-Kantorovich problem.

Definition 3. Let $\Pi(\pi_{Z^f}, \pi_{Z^a})$ denote the set of all possible couplings between π_{Z^f} and π_{Z^a}. A transference plan $\pi^*_{Z^{f:a}} \in \Pi(\pi_{Z^f}, \pi_{Z^a})$ is called the solution to the *Monge-Kantorovich problem* with cost function $c(z^f, z^a) = \|z^f - z^a\|^2$ if

$$\pi^*_{Z^{f:a}} = \arg \inf_{\pi_{Z^{f:a}} \in \Pi(\pi_{Z^f}, \pi_{Z^a})} \mathbb{E}[\|Z^f - Z^a\|^2]. \tag{2.27}$$

The associated functional $W(\pi_{Z^f}, \pi_{Z^a})$, defined by

$$W(\pi_{Z^f}, \pi_{Z^a})^2 = \mathbb{E}[\|Z^f - Z^a\|^2] \tag{2.28}$$

is called the *L^2-Wasserstein distance* between π_{Z^f} and π_{Z^a}.

Example 1. Let us consider the discrete set

$$\mathbb{Z} = \{z_1, z_2, \ldots, z_M\}, \qquad z_i \in \mathbb{R}, \tag{2.29}$$

and two probability vectors $\mathbb{P}(z_i) = 1/M$ and $\mathbb{P}(z_i) = w_i$, respectively, on $\mathbb{Z}$ with $w_i \geq 0$, $i = 1, \ldots, M$, and $\sum_i w_i = 1$. Any coupling between these two probability vectors is characterized by a matrix $T \in \mathbb{R}^{M \times M}$ such that its entries $t_{ij} = (T)_{ij}$ satisfy $t_{ij} \geq 0$ and

$$\sum_{i=1}^{M} t_{ij} = 1/M, \qquad \sum_{j=1}^{M} t_{ij} = w_i. \tag{2.30}$$

These matrices characterize the set of all couplings Π in the definition of the Monge-Kantorovich problem. Given a coupling T and the mean values

$$\bar{z}^f = \frac{1}{M} \sum_{i=1}^{M} z_i, \qquad \bar{z}^a = \sum_{i=1}^{M} w_i z_i,$$

the covariance between the associated discrete random variables $Z^f : \Omega \to \mathbb{Z}$ and $Z^a : \Omega \to \mathbb{Z}$ is defined by

$$\text{cov}(Z^f, Z^a) = \sum_{i,j=1}^{M} (z_i - \bar{z}^a) t_{ij} (z_j - \bar{z}^f). \tag{2.31}$$

The particular coupling defined by $t_{ij} = w_i/M$ leads to zero correlation between Z^f and Z^a. On the other hand, maximizing the correlation leads to a *linear transport problem* in the M^2 unknowns $\{t_{ij}\}$. More precisely, the unknowns t_{ij} have to satisfy the inequality constraints $t_{ij} \geq 0$, the equality constraints (2.30), and should minimize

$$J(T) = \sum_{i,j=1}^{M} t_{ij} |z_i - z_j|^2,$$

which is equivalent to maximizing (2.31). See Strang (1986) and Nocedal and Wright (2006) for an introduction to linear transport problems and, more generally, to linear programming.

We now return to continuous random variables and the desired coupling between forecast and analysis PDFs. The following theorem is an adaptation of a more general result on optimal couplings from Villani (2003).

Theorem 2. *If the forecast PDF π_{Z^f} has bounded second-order moments, then the optimal transference plan that solves the Monge-Kantorovich problem gives rise to a deterministic coupling with transport map*

$$Z^a = \nabla_z \phi(Z^f),$$

where $\phi : \mathbb{R}^{N_z} \to \mathbb{R}$ is a convex potential.

Below we sketch the basic line of arguments that lead to this theorem. We first introduce the *support* of a coupling $\pi_{Z^{f:a}}$ on $\mathbb{R}^{N_z} \times \mathbb{R}^{N_z}$ as the smallest closed set on which $\pi_{Z^{f:a}}$ is concentrated, *i.e.*,

$$\text{supp}(\pi_{Z^{f:a}}) := \bigcap \{S \subset \mathbb{R}^{N_z} \times \mathbb{R}^{N_z} : S \text{ closed and } \mu_{Z^{f:a}}(\mathbb{R}^{N_z} \times \mathbb{R}^{N_z} \setminus S) = 0\}$$

with the measure of $\mathbb{R}^{N_z} \times \mathbb{R}^{N_z} \setminus S$ defined by

$$\mu_{Z^{f:a}}(\mathbb{R}^{N_z} \times \mathbb{R}^{N_z} \setminus S) = \int_{\mathbb{R}^{N_z} \times \mathbb{R}^{N_z} \setminus S} \pi_{Z^{f:a}}(z^f, z^a) \, dz^f \, dz^a.$$

The support of $\pi_{Z^{f:a}}$ is called *cyclically monotone* if for every set of points $(z_i^f, z_i^a) \in$ supp $(\pi_{Z^{f:a}}) \subset \mathbb{R}^{N_z} \times \mathbb{R}^{N_z}$, $i = 1, \ldots, I$, and any permutation σ of $\{1, \ldots, I\}$ one has

$$\sum_{i=1}^{I} \|z_i^f - z_i^a\|^2 \leq \sum_{i=1}^{I} \|z_i^f - z_{\sigma(i)}^a\|^2. \tag{2.32}$$

Note that (2.32) is equivalent to

$$\sum_{i=1}^{I} (z_i^f)^T (z_{\sigma(i)}^a - z_i^a) \leq 0.$$

It can be shown that any coupling whose support is not cyclically monotone can be modified into another coupling with lower transport cost. Hence it follows that a solution $\pi_{Z^{f:a}}^*$ of the Monge-Kantorovich problem (2.27) has cyclically monotone support.

A fundamental theorem (*Rockafellar's theorem* (Villani, 2003)) of convex analysis now states that cyclically monotone sets $S \subset \mathbb{R}^{N_z} \times \mathbb{R}^{N_z}$ are contained in the *subdifferential* of a convex function $\phi : \mathbb{R}^{N_z} \to \mathbb{R}$. Here the subdifferential $\partial \phi$ of a convex function ϕ at a point $z \in \mathbb{R}^{N_z}$ is defined as the compact, non-empty and convex set of all $m \in \mathbb{R}^{N_z}$ such that

$$\phi(z') \geq \phi(z) + m(z' - z)$$

for all $z' \in \mathbb{R}^{N_z}$. We write $m \in \partial \phi(z)$. An optimal transport map is obtained whenever the convex potential ϕ for a given optimal coupling $\pi_{Z^{f:a}}^*$ is sufficiently regular in which case the subdifferential $\partial \phi(z)$ reduces to the classic gradient $\nabla_z \phi$ and $z^a = \nabla_z \phi(z^f)$. This regularity is ensured by the assumptions of the above theorem. See McCann (1995) and Villani (2003) for more details.

We summarize the *McKean optimal transportation approach* in the following definition.

Definition 4. Given a dynamic iteration (2.6) with PDF π_{Z^0} for the initial conditions and observations y_{obs}^n, $n = 1, \ldots, N$, the forecast $Z^{n,f}$ at iteration index $n > 0$ is defined by (2.22) and the analysis $Z^{n,a}$ by (2.25). The convex potential ϕ^n is the solution to the Monge-Kantorovich optimal transportation problem for coupling $\pi_{Z^{f,n}}$ and $\pi_{Z^{a,n}}$. The iteration is started at $n = 0$ with $Z^{0,a} = Z^0$.

The application of optimal transportation to Bayesian inference and data assimilation has first been discussed by Reich (2011), Moselhy and Marzouk (2012), and Reich and Cotter (2013).

In the following section we discuss data assimilation algorithms from a McKean optimal transportation perspective.

5 Linear ensemble transform methods

In this section, we discuss SMCMs, EnKFs, and the recently proposed (Reich, 2013a) ETPF from a coupling perspective. All three data assimilation methods have in common that they are based on an ensemble z_i^n, $i = 1, \ldots, M$, of model states. In the absence of observations the M ensemble members propagate independently under the model dynamics (2.6), *i.e.*, an analysis ensemble at time-level $n - 1$ gives rise to a forecast ensemble at time-level n via

$$z_i^{n,f} = \Psi(z_i^{n-1,a}), \qquad i = 1, \ldots, M.$$

The three methods differ in the way the forecast ensemble $\{z_i^{n,f}\}_{i=1}^M$ is transformed into an analysis ensemble $\{z_i^{n,a}\}_{i=1}^M$ under an observation y_{obs}^n. However, all three methods share a common mathematical structure which we outline next. We drop the iteration index in order to simplify the notation.

Definition 5. The class of *linear ensemble transform filters* (LETFs) is defined by

$$z_j^a = \sum_{i=1}^M z_i^f s_{ij}, \tag{2.33}$$

where the coefficients s_{ij} are the M^2 entries of a matrix $S \in \mathbb{R}^{M \times M}$.

The concept of LETFs is well established for EnKF formulations (Tippett et al., 2003; Evensen, 2006). It will be shown below that SMCMs and the ETPF also belong to the class of LETFs. In other words, these three methods differ only in the definition of the corresponding transform matrix S.

5.1 Sequential Monte Carlo methods (SMCMs)

A central building block of an SMCM is the *proposal density* $\pi_{\text{prop}}(z|z')$, which produces a *proposal ensemble* $\{z_i^p\}_{i=1}^M$ from the last analysis ensemble. Here we assume, for simplicity, that the proposal density is given by the model dynamics itself, *i.e.*,

$$\pi_{\text{prop}}(z|z') = \delta(z - \Psi(z')),$$

and then

$$z_i^p = z_i^f, \qquad i = 1, \ldots, M.$$

One associates with the proposal/forecast ensemble two discrete measures on

$$\mathbb{Z} = \{z_1^f, z_2^f, \ldots, z_M^f\}, \tag{2.34}$$

namely the uniform measure $\mathbb{P}(z_i^f) = 1/M$ and the non-uniform measure

$$\mathbb{P}(z_i^f) = w_i,$$

defined by the importance weights

$$w_i = \frac{\pi_Y(y_{\mathrm{obs}}|z_i^f)}{\sum_{j=1}^{M} \pi_Y(y_{\mathrm{obs}}|z_j^f)}. \tag{2.35}$$

The *sequential importance resampling* (SIR) filter (Gordon et al., 1993) resamples from the weighted forecast ensemble in order to produce a new equally weighted analysis ensemble $\{z_i^a\}$. Here we only consider SIR filter implementations with resampling performed after each data assimilation cycle.

An in-depth discussion of the SIR filter and more general SMCMs can be found, for example, in Doucet et al. (2001); Doucet and Johansen (2011). Here we focus on the coupling of discrete measures perspective of a resampling step. We first note that any resampling strategy effectively leads to a coupling of the uniform and the non-uniform measure on (2.34). As previously discussed, a coupling is defined by a matrix $T \in \mathbb{R}^{M \times M}$ such that $t_{ij} \geq 0$, and

$$\sum_{i=1}^{M} t_{ij} = 1/M, \qquad \sum_{j=1}^{M} t_{ij} = w_i. \tag{2.36}$$

The resampling strategy (2.20) leads to

$$t_{ij} = \frac{1}{M} \left(\epsilon w_j \delta_{ij} + (1 - \epsilon w_j) w_i \right)$$

with $\epsilon \geq 0$ chosen such that $\epsilon w_j \leq 1$ for all $j = 1, \ldots, M$. Monomial resampling corresponds to the special case $\epsilon = 0$, *i.e.* $t_{ij} = w_i/M$. The associated transformation matrix S in (2.33) is the realization of a random matrix with entries $s_{ij} \in \{0, 1\}$ such that each column of S contains exactly one entry $s_{ij} = 1$. Given a coupling T, the probability for the s_{ij} entry to be the one selected in the jth column is

$$\mathbb{P}(s_{ij} = 1) = M t_{ij}$$

and $M t_{ij} = w_i$ in case of monomial resampling. Any such resampling procedure based on a coupling matrix T leads to a consistent coupling for the underlying forecast and analysis probability measures as $M \to \infty$, which, however, is non-optimal in the sense of the Monge-Kantorovich problem (2.27). We refer to Bain and Crisan (2009) for resampling strategies which satisfy alternative optimality conditions.

We emphasize that the transformation matrix S of a SIR particle filter analysis step satisfies

$$\sum_{i=1}^{M} s_{ij} = 1 \qquad (2.37)$$

and

$$s_{ij} \in [0, 1]. \qquad (2.38)$$

In other words, each realization of the resampling step yields a Markov chain S. Furthermore, the weights $\hat{w}_i = M^{-1} \sum_{j=1}^{M} s_{ij}$ satisfy $\mathbb{E}[\hat{W}_i] = w_i$ and the analysis ensemble defined by $z_j^a = z_i^f$ if $s_{ij} = 1, j = 1, \ldots, M$, is contained in the convex hull of the forecast ensemble (2.34).

A forecast ensemble $\{z_i^f\}_{i=1}^{M}$ leads to the following estimator

$$\bar{z}^f = \frac{1}{M} \sum_{i=1}^{M} z_i^f$$

for the mean and

$$P_{zz}^f = \frac{1}{M-1} \sum_{i=1}^{M} (z_i^f - \bar{z}^f)(z_i^f - \bar{z}^f)^T$$

for the covariance matrix. In order to increase the *robustness* of a SIR particle filter one often augments the resampling step by the *particle rejuvenation step* (Pham, 2001)

$$z_j^a = z_i^f + \xi_j, \qquad (2.39)$$

where the ξ_j's are realizations of M independent and identically distributed Gaussian random variables $\mathrm{N}(0, h^2 P_{zz}^f)$ and $s_{ij} = 1$. Here $h > 0$ is the *rejuvenation parameter* which determines the magnitude of the stochastic perturbations. Rejuvenation helps to avoid the creation of identical analysis ensemble members which would remain identical under the deterministic model dynamics (2.6) for all times. Furthermore, rejuvenation can be used as a heuristic tool in order to compensate for model errors as encoded, for example, in the difference between (2.6) and (2.13).

In this paper we only discuss SMCMs which are based on the proposal step (2.4). Alternative proposal steps are possible and recent work on alternative implementations of SMCMs include van Leeuwen (2010), Chorin et al. (2010), Morzfeld et al. (2012), Morzfeld and Chorin (2012), van Leeuwen and Ades (2012), Reich (2013b).

5.2   Ensemble Kalman filter (EnKF)

The historically first version of the EnKF uses perturbed observations in order to transform a forecast ensemble into an analysis ensemble. The key requirement of any EnKF is that the transformation step is consistent with the classic Kalman update step in case the forecast and analysis PDFs are Gaussian. The, so-called, *EnKF with perturbed observations* is explicitly given by the simple formula

$$z_j^a = z_j^f - K(y_j^f + \xi_j - y_{\text{obs}}), \qquad j = 1, \ldots, M,$$

where $y_j^f = h(z_j^f)$, the ξ_j's are realizations of independent and identically distributed Gaussian random variables with mean zero and covariance matrix R, and K denotes the Kalman gain matrix, which in case of the EnKF is determined by the forecast ensemble as follows:

$$K = P_{zy}^f (P_{yy}^f + R)^{-1}$$

with empirical covariance matrices

$$P_{zy}^f = \frac{1}{M-1} \sum_{i=1}^{M} (z_i^f - \bar{z}^f)(y_i^f - \bar{y}^f)^T$$

and

$$P_{yy}^f = \frac{1}{M-1} \sum_{i=1}^{M} (y_i^f - \bar{y}^f)(y_i^f - \bar{y}^f)^T,$$

respectively. Here the ensemble mean in observation space is defined by

$$\bar{y}^f = \frac{1}{M} \sum_{i=1}^{M} y_i^f.$$

In order to shorten subsequent notations, we introduce the $N_y \times M$ matrix of ensemble deviations

$$A_y^f = [y_1^f - \bar{y}^f, y_2^f - \bar{y}^f, \ldots, y_M^f - \bar{y}^f]$$

in observation space and the $N_z \times M$ matrix of ensemble deviations

$$A_z^f = [z_1^f - \bar{z}^f, z_2^f - \bar{z}^f, \ldots, z_M^f - \bar{z}^f]$$

in state space, respectively. In terms of these ensemble deviation matrices, it holds that

$$P^f_{zz} = \frac{1}{M-1} A^f_z (A^f_z)^T \quad \text{and} \quad P^f_{zy} = \frac{1}{M-1} A^f_z (A^f_y)^T,$$

respectively.

It can be verified by explicit calculations that the EnKF with perturbed observations fits into the class of LETFs with

$$z^a_j = \sum_{i=1}^{M} z^f_i \left(\delta_{ij} - \frac{1}{M-1} (y^f_i - \bar{y}^f_i)^T (P^f_{yy} + R)^{-1} (y^f_j + \xi_j - y_{\text{obs}}) \right)$$

and, therefore,

$$s_{ij} = \delta_{ij} - \frac{1}{M-1} (y^f_i - \bar{y}^f)^T (P^f_{yy} + R)^{-1} (y^f_j + \xi_j - y_{\text{obs}}).$$

Here we have used that

$$\frac{1}{M-1} \sum_{i=1}^{M} (z^f_i - \bar{z}^f)(y^f_i - \bar{y}^f)^T = \frac{1}{M-1} \sum_{i=1}^{M} z^f_i (y^f_i - \bar{y}^f)^T.$$

We note that the transform matrix S is the realization of a random matrix. The class of *ensemble square root filters* (ESRF) leads instead to deterministic transformation matrices S. More precisely, an ESRF uses separate transformation steps for the ensemble mean $\bar{z}^f$ and the ensemble deviations $z^f_i - \bar{z}^f$. The mean is simply updated according to the classic Kalman formula, *i.e.*

$$\bar{z}^a = \bar{z}^f - K(\bar{y}^f - y_{\text{obs}}) \tag{2.40}$$

with the Kalman gain matrix defined as before.

Upon introducing the analysis matrix of ensemble deviations $A^a_z \in \mathbb{R}^{N_z \times M}$, one obtains

$$P^a_{zz} = \frac{1}{M-1} A^a_z (A^a_z)^T$$

$$= P^f_{zz} - K(P^f_{zy})^T = \frac{1}{M-1} A^f_z Q (A^f_z)^T \tag{2.41}$$

with the $M \times M$ matrix Q defined by

$$Q = I - \frac{1}{M-1} (A^f_y)^T (P^f_{yy} + R)^{-1} A^f_y.$$

Let us denote the matrix square root[8] of Q by D and its entries by d_{ij}.

---

[8]The matrix square root of a symmetric positive semi-definite matrix Q is the unique symmetric matrix D which satisfies $DD = Q$.

We note that $\sum_{i=1}^{M} d_{ij} = 1$ and it follows that

$$
\begin{aligned}
z_j^a &= \bar{z}^f - K(\bar{y}^f - y_{\text{obs}}) + \sum_{i=1}^{M}(z_i^f - \bar{z}^f)d_{ij} \\
&= \sum_{i=1}^{M} z_i^f \left(\frac{1}{M-1}(y_i^f - \bar{y}^f)^T (P_{yy}^f + R)^{-1}(y_{\text{obs}} - \bar{y}^f) + d_{ij} \right) \\
&= \sum_{i=1}^{M} z_i^f \left(\frac{1}{M-1}(y_i^f - \bar{y}^f)^T (P_{yy}^f + R)^{-1}(y_{\text{obs}} - \bar{y}^f) + d_{ij} \right).
\end{aligned}
\tag{2.42}
$$

The appropriate entries for the transformation matrix S of an ESRF can now be read off of (2.42). See Tippett et al. (2003); Wang et al. (2004); Livings et al. (2008); Ott et al. (2004); Nerger et al. (2012) and Evensen (2006) for further details on other ESRF implementations such as the ensemble adjustment Kalman filter. We mention in particular that an application of the Sherman-Morrison-Woodbury formula (Golub and van Loan, 1996) leads to the equivalent square root formula

$$
D = \left\{ I + \frac{1}{M-1}(A_y^f)^T R^{-1} A_y^f \right\}^{-1/2},
\tag{2.43}
$$

which avoids the need for inverting the $N_y \times N_y$ matrix $P_{yy}^f + R$, which is desirable whenever $N_y \gg M$. Furthermore, using the equivalent Kalman gain matrix representation

$$
K = P_{zy}^a R^{-1},
$$

the Kalman update formula (2.40) for the mean becomes

$$
\begin{aligned}
\bar{z}^a &= \bar{z}^f - P_{zy}^a R^{-1}(\bar{y}^f - y_{\text{obs}}) \\
&= \bar{z}^f + \frac{1}{M-1} A_z^f Q(A_y^f)^T R^{-1}(y_{\text{obs}} - \bar{y}^f).
\end{aligned}
$$

This reformulation gives rise to

$$
s_{ij} = \frac{1}{M-1} q_{ij}(y_j^f - \bar{y}^f) R^{-1}(y_{\text{obs}} - \bar{y}^f) + d_{ij},
\tag{2.44}
$$

which forms the basis of the *local ensemble transform Kalman filter* (LETKF) (Ott et al., 2004; Hunt et al., 2007) to be discussed in more detail in Section 6.

We mention that the EnKF with perturbed observations or an ESRF implementation leads to transformation matrices S which satisfy (2.37) but the entries s_{ij} can take positive as well as negative values. This can be problematic in case the state variable z should be non-negative. Then it is possible that a forecast ensemble $z_i^f \geq 0$, $i = 1, \ldots, M$, is transformed into an analysis z_i^a, which contains negative entries. See Janjić et al. (2014) for modifications to EnKF type algorithms in order to preserve positivity.

One can discuss the various EnKF formulations from an optimal transportation perspective. Here the coupling is between two Gaussian distributions; the forecast PDF $N(\bar{z}^f, P_{zz}^f)$ and analysis PDF $N(\bar{z}^a, P_{zz}^a)$, respectively, with the analysis mean given by (2.40) and the analysis covariance matrix by (2.41). We know that the optimal coupling must be of the form

$$z^a = \nabla_z \phi(z^f)$$

and, in case of Gaussian PDFs, the convex potential $\phi : \mathbb{R}^{N_z} \to \mathbb{R}$ is furthermore bilinear, *i.e.*,

$$\phi(z) = b^T z + \frac{1}{2} z^T A z$$

with the vector b and the matrix A appropriately defined. The choice

$$z^a = b + A z^f = \bar{z}^a + A(z^f - \bar{z}^f)$$

leads to

$$b = \bar{z}^a - A \bar{z}^f$$

for the vector $b \in \mathbb{R}^{N_z}$. The matrix $A \in \mathbb{R}^{N_z \times N_z}$ then needs to satisfy

$$P_{zz}^a = A P_{zz}^f A^T.$$

The optimal, in the sense of Monge-Kantorovich with cost function $c(z^f, z^a) = \|z^f - z^a\|^2$, matrix A is given by

$$A = (P_{zz}^a)^{1/2} \left[(P_{zz}^a)^{1/2} P_{zz}^f (P_{zz}^a)^{1/2} \right]^{-1/2} (P_{zz}^a)^{1/2}.$$

See Olkin and Pukelsheim (1982). An efficient implementation of this optimal coupling in the context of ESRFs has been discussed in Reich and Cotter (2013). The essential idea is to replace the matrix square root of P_{zz}^a by the analysis matrix of ensemble deviations $A_z^a = A_z^f D$ scaled by $1/\sqrt{M-1}$.

Note that different cost functions $c(z^f, z^a)$ lead to different solutions to the associated Monge-Kantorovich problem (2.27). Of particular interest is the weighted inner product

$$c(z^f, z^a) = \left((z^f - z^a)^T B^{-1} (z^f - z^a) \right)^2$$

for an appropriate positive definite matrix $B \in \mathbb{R}^{N_z \times N_z}$ (Reich and Cotter, 2013).

As for SMCMs particle rejuvenation can be applied to the analysis from an EnKF or ESRF. However, the more popular method for increasing the robustness of EnKFs is to apply *multiplicative ensemble inflation*

$$z_i^f \to \bar{z}^f + \alpha(z_i^f - \bar{z}^f), \qquad \alpha \geq 1, \tag{2.45}$$

to the forecast ensemble prior to the application of an EnKF or ESRF. The parameter α is called the inflation factor. An adaptive strategy for determining the factor α has, for example, been proposed by Anderson (2007); Miyoshi (2011). The inflation factor α can formally be related to the rejuvenation parameter h in (2.39) through

$$\alpha = \sqrt{1 + h^2}.$$

This relation becomes exact as $M \to \infty$.

We mention that the *rank histogram filter* of Anderson (2010), which uses a nonlinear filter in observation space and linear regression from observation onto state space, also fits into the framework of the LETFs. See Reich and Cotter (2015) for more details. The *nonlinear ensemble adjustment filter* of Lei and Bickel (2011), on the other hand, falls outside the class of LETFs.

5.3   Ensemble transform particle filter (ETPF)

We now return to the SIR filter described in Section 5.1. Recall that a SIR filter relies on importance resampling which we have interpreted as a coupling between the uniform measure on (2.34) and the measure defined by (2.35). Any coupling is characterized by a matrix T such that its entries are non-negative and (2.36) hold.

Definition 6. The ETPF is based on choosing the T which minimizes

$$J(T) = \sum_{i,j=1}^{M} t_{ij} \|z_i^f - z_j^f\|^2 \tag{2.46}$$

subject to (2.36) and $t_{ij} \geq 0$. Let us denote the minimizer by T^*. Then the transform matrix S of an ETPF is defined by

$$S = MT^*,$$

which satisfies (2.37) and (2.38).

Let us give a geometric interpretation of the ETPF transformation step. Since T^* from Definition 6 provides an optimal coupling, Rockafellar's theorem implies the existence of a convex potential $\phi_M : \mathbb{R}^{N_z} \to \mathbb{R}$ such that

$$z_i^f \in \partial\phi_M(z_j^f) \quad \text{for all} \quad i \in \mathscr{I}_j := \{i' \in \{1,\dots,M\} : t_{i'j}^* > 0\},$$

$j = 1, \ldots, M$. In fact, ϕ_M can be chosen to be piecewise affine and a constructive formula can be found in Villani (2003). The ETPF transformation step

$$z_j^a = M \sum_{i=1}^{M} z_i^f t_{ij}^* = \sum_{i=1}^{M} z_i^f s_{ij} \qquad (2.47)$$

corresponds to a particular selection from the linear space $\partial \phi_M(z_j^f)$, $j = 1, \ldots, M$; namely the expectation value of the discrete random variable

$$Z_j^a : \Omega \rightarrow \{z_1^f, z_2^f, \ldots, z_M^f\}$$

with probabilities $\mathbb{P}(z_i^f) = s_{ij}$, $i = 1, \ldots, M$. Hence it holds that

$$\bar{z}^a := \frac{1}{M} \sum_{j=1}^{M} z_j^a = \sum_{i=1}^{M} w_i z_i^f.$$

See Reich (2013a) for more details, where it has also been shown that the potentials ϕ_M converge to the solution of the underlying continuous Monge-Kantorovich problem as the ensemble size M approaches infinity.

It should be noted that standard algorithms for finding the minimizer of (2.46) suffer from a $\mathcal{O}(M^3 \log M)$ computational complexity. This complexity has been reduced to $\mathcal{O}(M^2 \log M)$ by Pele and Werman (2009). There are also fast iterative methods for finding approximate minimizers of (2.46) using the *Sinkhorn distance* (Cuturi, 2013).

The particle rejuvenation step (2.39) for SMCMs can be extended to the ETPF as follows:

$$z_j^a = \sum_{i=1}^{M} z_i^f s_{ij} + \xi_j, \qquad j = 1, \ldots, M. \qquad (2.48)$$

As before the ξ_j's are realizations of M independent Gaussian random variables with mean zero and appropriate covariance matrices P_j^a. We use $P_j^a = h^2 P_{zz}^f$ with rejuvenation parameter $h > 0$ for the numerical experiments conducted in this paper. Another possibility would be to locally estimate P_j^a from the coupling matrix T^*, *i.e.*,

$$P_j^a = \sum_{i=1}^{M} s_{ij}(z_i^f - \bar{z}_j^a)(z_i^f - \bar{z}_j^a)^T$$

with mean $\bar{z}_j^a = \sum_{i=1}^{M} s_{ij} z_i^f$.

5.4 Quasi-Monte Carlo (QMC) convergence

The expected rate of convergence for standard Monte Carlo methods is $M^{-1/2}$ where M denotes the ensemble size. QMC methods have an upper bound of $\log(M)^d M^{-1}$ where d stands for the dimension (Caflisch, 1988). For the purpose of this paper, $d = N_z$. Unlike Monte Carlo methods QMC methods also depend on the dimension of the space which implies a better performance for small N_z or/and large M. However, in practice QMC methods perform considerably better than the theoretical bound for the convergence rate and outperform Monte Carlo methods even for small ensemble sizes and in very high dimensional models. The latter may be explained by the concept of *effective dimension* introduced by Caflisch et al. (1997).

The following simulation investigates the convergence rate of the estimators for the first and second moment of the posterior distribution after applying a single analysis step of a SIR particle filter and an ETPF. The prior is chosen to be a uniform distribution on the unit square and the sum of both components is observed with additive noise drawn from a centered Gaussian distribution with variance equal to two. Reference values for the posterior moments are generated using Monte Carlo importance sampling with sample size $M = 2^{26}$. QMC samples of different sizes are drawn from the prior distribution and a single residual resampling step is compared to a single transformation step using an optimal coupling T^*. Figure 1 shows the root mean square errors (RMSEs) of the different posterior estimates with respect to their reference values. We find that the transform method preserves the optimal M^{-1} convergence rate of the prior QMC samples while resampling reduces the convergence rate to the $M^{-1/2}$.

We mention that replacing the deterministic transformation step in (2.47) by drawing ensemble member j from the prior ensemble according to the weights given by the j-th column of S leads to a stochastic version of the ETPF. This variant, despite being stochastic like the importance resampling step, results again in a QMC convergence rate.

6 Spatially extended dynamical systems and localization

Let us start this section with a simple thought experiment on the curse of dimensionality. Consider a state space of dimension $N_z = 100$ and a prior Gaussian distribution with mean zero and covariance matrix $P^f = I$. The reference solution is $z_{\text{ref}} = 0$ and we observe every component of the state vector subject to independent measurement errors with mean zero and variance $R = 0.16$. If one applies a single importance resampling step to this problem with ensemble size $M = 10$, one finds that the effective sample size collapses to $M_{\text{eff}} \approx 1$ and the resulting analysis ensemble is unable to recover the reference solution. However, one also quickly realizes that the stated problem can be decomposed into N_z independent data assimilation problems in each component of the state vector alone. If importance

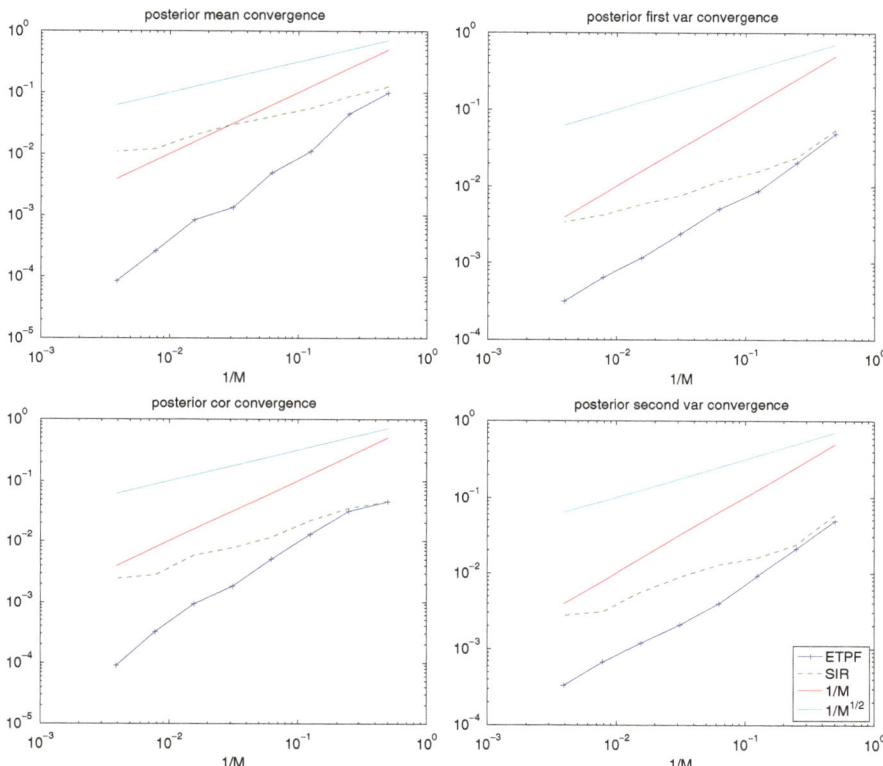

Fig. 1 RMSEs of estimates for the posterior mean, variances (var), and correlation (cor) using importance resampling (SIR) and optimal transformations (ETPF) plotted on a log-log scale as a function of ensemble sizes M.

resampling is now performed in each component of the state vector independently, then the effective sample size for each of the N_z analysis problems remains close to $M = 10$ and the reference solution can be recovered from the given set of observations. This is the idea of *localization*. Note that localization has increased the total sample size to $M \times N_z = 1000$ for this problem!

We now formally extend LETFs to spatially extended systems which may be viewed as an infinite-dimensional dynamical system (Robinson, 2001) and formulate an appropriate localization strategy. Consider the linear advection equation

$$u_t + u_x = 0$$

as a simple example for such a scenario. If $u_0(x)$ denotes the solution at time $t = 0$, then

$$u(x, t) = u_0(x + t)$$

is the solution of the linear advection equation for all $t \geq 0$. Given a time-increment $\Delta t > 0$, the associated dynamical system maps a function $u(x)$ into $u(x + \Delta t)$. A finite-dimensional dynamical system is obtained by discretizing in space with mesh-size $\Delta x > 0$. For example, the Box scheme (Morton and Mayers, 2005) leads to

$$\frac{u_j^{k+1} + u_{j+1}^{k+1} - u_j^k - u_{j+1}^k}{2\Delta t} + \frac{u_{j+1}^{k+1} + u_{j+1}^k - u_j^{k+1} - u_j^k}{2\Delta x} = 0$$

and, for J spatial grid points, the state vector at $t_k = k\Delta t$ becomes

$$z^k = (u_1^k, u_2^k, \dots, u_J^k)^T \in \mathbb{R}^J.$$

We may take the formal limit $J \to \infty$ and $\Delta x \to 0$ in order to return to functions $z^k(x)$. The dynamical system (2.6) is then defined as the map that propagates such functions (or their finite-difference approximations) from one observation instance to the next in accordance with the specified numerical method. Here we assume that observations are taken in intervals of $\Delta t_{\text{obs}} = N_{\text{out}} \Delta t$ with $N_{\text{out}} \geq 1$ a fixed integer. The index $n \geq 1$ in (2.6) is the counter for those observation instances.

In other words, forecast or analysis ensemble members, $z^{f/a}(x)$, now become functions of $x \in \mathbb{R}$, belong to some appropriate function space $\mathscr{H}$, and the dynamical system (2.6) is formally replaced by a map or evolution equation on $\mathscr{H}$ (Robinson, 2001). For simplicity of exposition we assume periodic boundary conditions, i.e., $z(x) = z(x + L)$ for some appropriate $L > 0$.

The curse of dimensionality (Bengtsson et al., 2008) implies that, generally speaking, none of the LETFs discussed so far is suitable for data assimilation of spatially extended systems. In order to overcome this situation, we now discuss the concept of *localization* as first introduced in Houtekamer and Mitchell (2001, 2005) for EnKFs. While we will focus on a particular localization, called *R-localization*, suggested by Hunt et al. (2007), our methodology can be extended to *B-localization* as proposed by Hamill et al. (2001).

In the context of the LETFs R-localization amounts to modifying (2.33) to

$$z_j^a(x) = \sum_{i=1}^M z_i^f(x) s_{ij}(x),$$

where the associated transform matrices $S(x) \in \mathbb{R}^M \times \mathbb{R}^M$ depend now on the spatial location $x \in [0, L]$. It is crucial that the transformation matrices $S(x)$ are sufficiently smooth in x in order to produce analysis ensembles with sufficient regularity for the evolution problem under consideration and, in particular $z_j^a \in \mathscr{H}$. In case of an SMCM with importance resampling, the resulting $S(x)$ would, in general, not even be continuous for almost all $x \in [0, L)$. Hence we only discuss localization for the ESRF and the ETPF.

Let us, for simplicity, assume that the forward operator $h : \mathcal{H} \to \mathbb{R}^{N_y}$ for the observations y_{obs} is defined by

$$h_k(z) = z(\mathsf{x}_k), \qquad k = 1, \ldots, N_y.$$

Here the $\mathsf{x}_k \in [0, L)$ denote the spatial location at which the observation is taken. The measurement errors are Gaussian with mean zero and covariance matrix $R \in \mathbb{R}^{N_y \times N_y}$. We assume for simplicity that R is diagonal.

In the sequel we assume that $z(x)$ has been extended to $x \in \mathbb{R}$ by periodic extension from $x \in [0, L)$ and introduce the time-averaged and normalized *spatial correlation function*

$$C(x, s) := \frac{\sum_{n=0}^{N} z^n(x + s) z^n(x)}{\sum_{n=0}^{N} (z^n(x))^2} \tag{2.49}$$

for $x \in [0, L)$ and $s \in [-L/2, L/2)$. Here we have assumed that the underlying solution process is stationary ergodic. In case of spatial homogeneity the spatial correlation function becomes furthermore independent of x for N sufficiently large.

We also introduce a localization kernel $\mathscr{K}(x, x'; r_{\mathrm{loc}})$ in order to define R-localization for an ESRF and the ETPF. The localization kernel can be as simple as

$$\mathscr{K}(x, x; r_{\mathrm{loc}}) = \begin{cases} 1 - \frac{1}{2}s & \text{for } s \leq 2, \\ 0 & \text{else,} \end{cases} \tag{2.50}$$

with

$$s := \frac{\min\{|x - x' - L|, |x - x'|, |x - x' + L|\}}{r_{\mathrm{loc}}} \geq 0,$$

or a higher-order polynomial such as

$$\mathscr{K}(x, x'; r_{\mathrm{loc}}) = \begin{cases} 1 - \frac{5}{3}s^2 + \frac{5}{8}s^3 + \frac{1}{2}s^4 - \frac{1}{4}s^5 & \text{for } s \leq 1, \\ -\frac{2}{3}s^{-1} + 4 - 5s + \frac{5}{3}s^2 + \frac{5}{8}s^3 - \frac{1}{2}s^4 + \frac{1}{12}s^5 & \text{for } 1 \leq s \leq 2, \\ 0 & \text{else.} \end{cases} \tag{2.51}$$

See Gaspari and Cohn (1999).

In order to compute the transformation matrix $S(x)$ for given x, we modify the kth diagonal entry in the measurement error covariance matrix $R \in \mathbb{R}^{N_y \times N_y}$ and define

$$\frac{1}{\tilde{r}_{kk}(x)} := \frac{\mathscr{K}(x, \mathsf{x}_k; r_{\mathrm{loc},R})}{r_{kk}} \tag{2.52}$$

for $k = 1, \ldots, N_y$. Given a localization radius $r_{\mathrm{loc},R} > 0$, this results in a matrix $\tilde{R}^{-1}(x)$ which replaces the R^{-1} in an ESRF and the ETPF.

More specifically, the LETKF is based on the following modifications to the ESRF. First one replaces (2.43) by

$$Q(x) = \left\{ I + \frac{1}{M - 1} (A_y^f)^T \tilde{R}^{-1}(x) A_y^f \right\}^{-1}$$

and defines $D(x) = Q(x)^{1/2}$. Finally the localized transformation matrix $S(x)$ is given by

$$s_{ij}(x) = \frac{1}{M-1} q_{ij}(x)(y_j^f - \bar{y}^f)\tilde{R}^{-1}(x)(y_{\text{obs}} - \bar{y}^f) + d_{ij}(x), \qquad (2.53)$$

which replaces (2.44). We mention that Anderson (2012) discusses practical methods for choosing the localization radius $r_{\text{loc},R}$ for EnKFs.

In order to extend the concept of R-localization to the ETPF, we also define the localized cost function

$$c_x(z^f, z^a) = \int_0^L \mathcal{K}(x, x'; r_{\text{loc},c}) \|z^f(x') - z^a(x')\|^2 \mathrm{d}x' \qquad (2.54)$$

with a localization radius $r_{\text{loc},c} \geq 0$, which can be chosen independently from the localization radius for the measurement error covariance matrix R.

The ETPF with R-localization can now be implemented as follows. At each spatial location $x \in [0, L)$ one determines the desired transformation matrix $S(x)$ by first computing the weights

$$w_i \propto e^{-\frac{1}{2}(h(z_i^f)-y_{\text{obs}})^T \tilde{R}^{-1}(x)(h(z_i^f)-y_{\text{obs}})} \qquad (2.55)$$

and then minimizing the cost function

$$J(T) = \sum_{i,j=1}^M c_x(z_i^f, z_j^f) t_{ij} \qquad (2.56)$$

over all admissible couplings. One finally sets $S(x) = MT^*$.

As discussed earlier any infinite-dimensional evolution equation such as the linear advection equation will be truncated in practice to a computational grid $x_j = j\Delta x$. The transform matrices $S(x)$ need then to be computed for each grid point only and the integral in (2.54) is replaced by a simple Riemann sum.

We mention that alternative filtering strategies for spatio-temporal processes have been proposed by Majda and Harlim (2012) in the context of turbulent systems. One of their strategies is to perform localization in spectral space in case of regularly spaced observations. Another spatial localization strategy for particle filters can be found in Rebeschini and van Handel (2013).

7 Applications

In this section we present some numerical results comparing the different LETFs for the chaotic Lorenz-63 (Lorenz, 1963) and Lorenz-96 (Lorenz, 1996) models. While the highly nonlinear Lorenz-63 model can be used to investigate the behavior of different DA algorithms for strongly non-Gaussian distributions, the forty dimensional Lorenz-96 model is a prototype "spatially extended" system which demonstrates the need for localization in order to achieve skillful filter results for moderate ensemble sizes. We begin with the Lorenz-63 model.

We mention that theoretical results on the long time behavior of filtering algorithms for chaotic systems, such as the Lorenz-63 model, have been obtained, for example, by González-Tokman and Hunt (2013) and Law et al. (2013).

7.1 *Lorenz-63 model*

The Lorenz-63 model is given by the differential equation (2.7) with state variable $z = (x, y, z)^T \in \mathbb{R}^3$, right-hand side

$$f(z) = \begin{pmatrix} \sigma(y - x) \\ x(\rho - z) - y \\ xy - \beta z \end{pmatrix},$$

and parameter values $\sigma = 10$, $\rho = 28$, and $\beta = 8/3$. The resulting ODE (2.7) is discretized in time by the implicit midpoint method (Ascher, 2008), *i.e.*,

$$z^{n+1} = z^n + \Delta t f(z^{n+1/2}), \qquad z^{n+1/2} = \frac{1}{2}(z^{n+1} + z^n) \tag{2.57}$$

with step-size $\Delta t = 0.01$. Let us abbreviate the resulting map by Ψ_{IM}. Then the dynamical system (2.6) is defined as

$$\Psi = \Psi_{\text{IM}}^{[12]}.$$

In other words observations are assimilated every 12 time-steps. We only observe the x variable with a Gaussian measurement error of variance $R = 8$.

We used different ensemble sizes from 10 to 80 as well as different inflation factors ranging from 1.0 to 1.12 by increments of 0.02 for the EnKF and rejuvenation parameters ranging from 0 to 0.4 by increments of 0.04 for the ETPF. Note that a rejuvenation parameter of $h = 0.4$ corresponds to an inflation factor $\alpha = \sqrt{1 + h^2} \approx 1.0770$.

The following variant of the ETPF with localized cost function has also been implemented. We first compute the importance weights w_i of a given observation. Then each component of the state vector is updated using only the distance in that component in the cost function $J(T)$. For example, the x_i^f components of the forecast ensemble members $z_i^f = (x_i^f, y_i^f, z_i^f)^T$, $i = 1, \ldots, M$, are updated according to

$$x_i^a = M \sum_{i=1}^{M} x_i^f t_{ij}^*$$

with the coefficients $t_{ij}^* \geq 0$ minimizing the cost function

$$J(T) = \sum_{i,j=1}^{M} t_{ij} |x_i^f - x_j^f|^2$$

subject to (2.36). We use ETPF_R0 as the shorthand form for this method. This variant of the ETPF is of special interest from a computational point of view since the linear transport problem in $\mathbb{R}^3$ reduces to three simple one-dimensional problems.

The model is run over $N = 20,000$ assimilation steps after discarding 200 steps to lose the influence of the initial conditions. The resulting *root-mean-square errors* averaged over time (RMSEs)

$$\text{RMSE} = \frac{1}{N} \sum_{n=1}^{N} \sqrt{\|\bar{z}^{n,a} - z_{\text{ref}}^n\|^2}$$

are reported in Fig. 2 a)–c). We dropped the results for the ETPF and ETPF_R0 with ensemble size $M = 10$ as they indicated strong divergence. We see that the

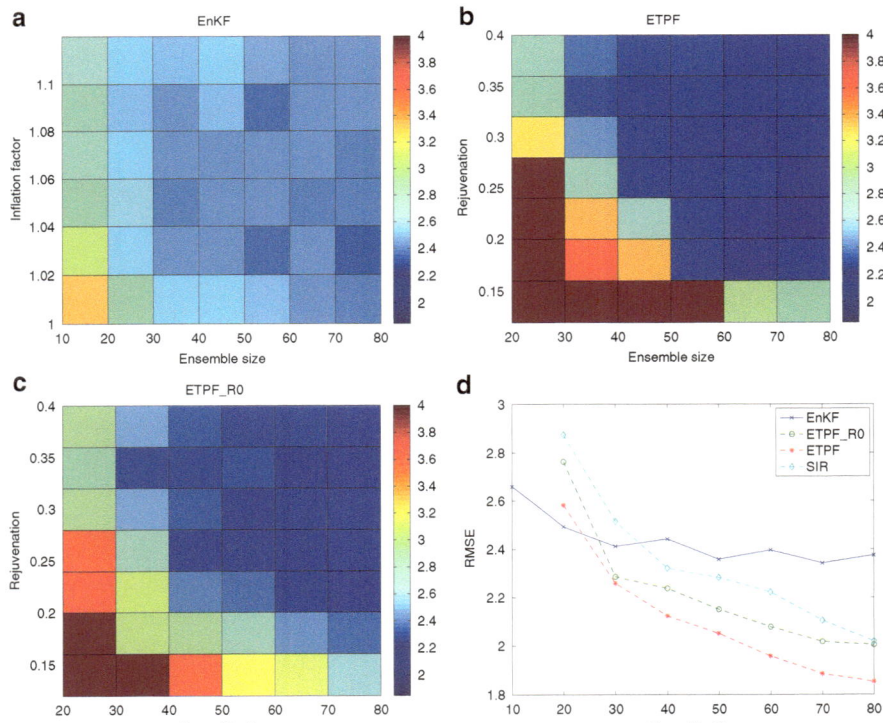

Fig. 2 a)-c): Heatmaps showing the RMSEs for different parameters for the EnKF, ETPF, and ETPF_R0 for the Lorenz-63 model. d): RMSEs for different ensemble sizes using 'optimal' inflation factors and rejuvenation.

EnKF produces stable results while the other filters are more sensitive to different choices for the rejuvenation parameter. However, with increasing ensemble size and 'optimal' choice of parameters the ETPF and the ETPF_R0 outperform the EnKF which reflects the biasedness of the EnKF.

Figure 2 d) shows the RMSEs for each ensemble size using the parameters that yield the lowest RMSE. Here we see again that the stability of the EnKF leads to good results even for very small ensemble sizes. The downside is also evident: While the ETPFs fail to track the reference solution as well as the EnKF for very small ensemble sizes a small increase leads to much lower RMSEs. The asymptotic consistent ETPF outperforms the ETPF_R0 for large ensemble sizes but is less stable otherwise. We also included RMSEs for the SIR filter with rejuvenation parameters chosen from the same range of values as for the ETPFs. Although not shown here, this range seems to cover the 'optimal' choice for the rejuvenation parameter. The comparison with the EnKF is as expected: for small ensemble sizes the SIR performs worse but beats the EnKF for larger ensemble sizes due to its asymptotic consistency. However, the equally consistent ETPF yields lower RMSEs throughout for the ensemble sizes considered here. Interestingly, the SIR only catches up with the inconsistent but computationally cheap ETPF_R0 for the largest ensemble size in this experiment. We mention that the RMSE drops to around 1.4 with the SIR filter with an ensemble size of 1000.

At this point we note that the computational burden increases considerably for the ETPF for larger ensemble sizes due to the need of solving increasingly large linear transport problems. See the discussion from Section 5.3.

7.2 Lorenz-96 model

Given a periodic domain $x \in [0, L]$ and N_z equally spaced grid-points $x_j = j\Delta x$, $\Delta x = L/N_z$, we denote by u_j the approximation to $z(x)$ at the grid points $x_j, j = 1, \ldots, N_z$. The following system of differential equations

$$\frac{\mathrm{d}u_j}{\mathrm{d}t} = -\frac{u_{j-1}u_{j+1} - u_{j-2}u_{j-1}}{3\Delta x} - u_j + F, \qquad j = 1, \ldots, 40, \qquad (2.58)$$

is due to Lorenz (1996) and is called the Lorenz-96 model. We set $F = 8$ and apply periodic boundary conditions $u_j = u_{j+40}$. The state variable is defined by $z = (u_1, \ldots, u_{40})^T \in \mathbb{R}^{40}$. The Lorenz-96 model (2.58) can be seen as a coarse spatial approximation to the PDE

$$\frac{\partial u}{\partial t} = -\frac{1}{2}\frac{\partial (u)^2}{\partial x} - u + F, \qquad x \in [0, 40/3],$$

with mesh-size $\Delta x = 1/3$ and $N_z = 40$ grid points. The implicit midpoint method (2.57) is used with a step-size of $\Delta t = 0.005$ to discretize the differential equations (2.58) in time. Observations are assimilated every 22 time-steps and we observe every other grid point with a Gaussian measurement error of variance $R = 8$. The large assimilation interval and variance of the measurement error are chosen because of a desired non-Gaussian ensemble distribution.

We used ensemble sizes from 10 to 80, inflation factors from 1.0 to 1.12 with increments of 0.02 for the EnKF and rejuvenation parameters between 0 and 0.4 with increments of 0.05 for the ETPFs.

As mentioned before, localization is required and we take (2.51) as our localization kernel. For each value of M we fixed a localization radius $r_{\text{loc},R}$ in (2.52). The particular choices can be read off of the following table:

M	10	20	30	40	50	60	70	80
$r_{\text{loc},R}^{EnKF}$	2	4	6	6	7	7	8	8
$r_{\text{loc},R}^{ETPF}$	1	2	3	4	5	6	6	6

These values have been found by trial and error and we do not claim that these values are by any means 'optimal'.

As for localization of the cost function (2.56) for the ETPF we used the same kernel as for the measurement error and implemented different versions of the localized ETPF which differ in the choice of the localization radius: ETPF_R1 corresponds to the choice of $r_{\text{loc},c} = 1$ and ETPF_R2 is used for the ETPF with $r_{\text{loc},c} = 2$. As before we denote the computationally cheap version with cost function $c_{x_j}(z^f, z^a) = |u_j^f - u_j^a|^2$ at grid point x_j by ETPF_R0.

The localization kernel and the localization radii $r_{\text{loc},c}$ are not chosen by any optimality criterion but rather by convenience and simplicity. A better kernel or localization radii may be derived from looking at the time averaged spatial correlation coefficients (2.49) as shown in Fig. 3. Our kernel gives higher weights

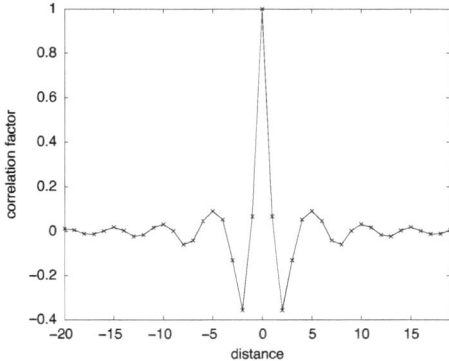

Fig. 3 Time averaged spatial correlation between solution components depending on their distance.

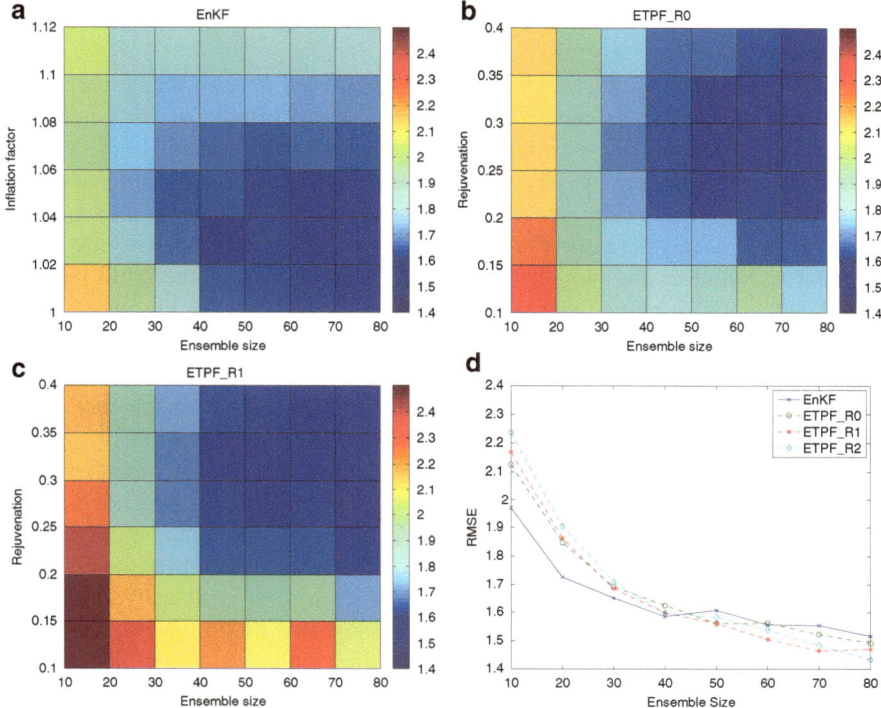

Fig. 4 a)-c): Heatmaps showing the RMSEs for different parameters for the EnKF, ETPF_R0, and ETPF_R1 for the Lorenz-96 model. d): RMSEs for different ensemble sizes using 'optimal' inflation factors and rejuvenation.

to components closer to the one to be updated, even though the correlation with the immediate neighbor is relatively low.

The model is run over $N = 10,000$ assimilation steps after discarding 500 steps to lose the influence of the initial conditions. The resulting time averaged RMSEs are displayed in Fig. 4. We dropped the results for the smallest rejuvenation parameters as the filters showed strong divergence. Similar to the results for the Lorenz-63 model the EnKF shows the most stable overall performance for various parameters but fails to keep up with the ETPFs for higher ensemble sizes, though the difference between the different filters is much smaller than for the Lorenz-63 system. This is no surprise since the Lorenz-96 system does not have the highly non-linear dynamics of the Lorenz-63 system which causes the involved distributions to be strongly non-Gaussian. The important point here is that the ETPF as a particle filter is able to compete with the EnKF even for small ensemble sizes. Traditionally, high dimensional systems required very high ensemble sizes for particle filters to perform reasonably well. Hundreds of particles are necessary for the SIR to be even close to the true state.

8 Historical comments

The notion of data assimilation has been coined in the field of meteorology and more widely in the geosciences to collectively denote techniques for combining computational models and physical observations in order to estimate the current state of the atmosphere or any other geophysical process. The perhaps first occurrence of the concept of data assimilation can be found in the work of Richardson (1922), where observational data needed to be interpolated onto a grid in order to initialize the computational forecast process. With the rapid increase in computational resolution starting in the 1960s, it became quickly necessary to replace simple data interpolation by an optimal combination of first guess estimates and observations. This gave rise to techniques such as the *successive correction method*, *nudging*, *optimal interpolation*, and *variational least square techniques* (3D-Var and 4D-Var). See Daley (1993); Kalnay (2002) for more details.

Leith (1974) proposed *ensemble* (or Monte Carlo) *forecasting* as an alternative to conventional single forecasts. However ensemble forecasting did not become operational before 1993 due to limited computer resources (Kalnay, 2002). The availability of ensemble forecasts subsequently lead to the invention of the EnKF by Evensen (1994) with a later correction by Burgers et al. (1998) and many subsequent developments, which have been summarized in Evensen (2006). We mention that the analysis step of an EnKF with perturbed observations is closely related to a method now called *randomized likelihood method* (Kitanidis, 1995; Oliver, 1996; Oliver et al., 1996).

In a completely independent line of research the problem of optimal estimation of stochastic processes from data has led to the theory of *filtering* and *smoothing*, which started with the work of Wiener (1948). The state space approach to filtering of linear systems gave rise to the celebrated *Kalman filter* and more generally to the *stochastic PDE formulations* of Zakai and Kushner-Stratonovitch in case of continuous-time filtering. See Jazwinski (1970) for the theoretical developments up to 1970. Monte Carlo techniques were first introduced to the filtering problem by Handschin and Mayne (1969), but it was not until the work of Gordon et al. (1993) that the SMCM became widely used (Doucet et al., 2001). The *McKean interacting particle approach* to SMCMs has been pioneered by del Moral (2004). The theory of particle filters for time-continuous filtering problems is summarized in Bain and Crisan (2009).

Standard SMCMs suffer from the *curse of dimensionality* in that the necessary number of ensemble members M increases exponentially with the dimension N_z of state space (Bengtsson et al., 2008). This limitation has prevented SMCMs from being used in meteorology and the geosciences. On the other hand, it is known that EnKFs lead to inconsistent estimates which is problematic when multimodal forecast distributions are to be expected. Current research work is therefore focused on a theoretical understanding of EnKFs and related sequential assimilation techniques (see, for example, González-Tokman and Hunt (2013); Law et al. (2013)), extensions of particle filters/SMCMs to PDE models (see, for example, Morzfeld and Chorin (2012); van Leeuwen and Ades (2012); Beskov et al. (2013); Metref et al. (2013)), and Bayesian inference on function spaces (see, for example, Stuart (2010); Cotter et al. (2009); Dashti et al. (2013)) and hybrid variational methods such as by, for example, Bonavita et al. (2012); Clayton et al. (2013).

A historical account of optimal transportation can be found in Villani (2009). The work of McCann (1995) provides the theoretical link between the classic linear assignment problem and the Monge-Kantorovich problem of coupling PDFs. The ETPF is a computational procedure for approximating such couplings using importance sampling and linear transport instead.

9 Summary and Outlook

We have discussed various ensemble/particle-based algorithms for sequential data assimilation in the context of LETFs. Our starting point was the McKean interpretation of Feynman-Kac formulae. The McKean approach requires a coupling of measures which can be discussed in the context of optimal transportation. This approach leads to the ETPF when applied in the context of SMCMs. We have furthermore discussed extensions of LETFs to spatially extended systems in form of R-localization.

The presented work can be continued along several lines. First, one may replace the empirical forecast measure

$$\pi_{\text{emp}}^{f}(z) := \frac{1}{M} \sum_{i=1}^{M} \delta(z - z_i^f), \tag{2.59}$$

which forms the basis of SMCMs and the ETPF, by a *Gaussian mixture*

$$\pi_{\text{GM}}^{f}(z) := \frac{1}{M} \sum_{i=1}^{M} \text{n}(z; z_i^f, B), \tag{2.60}$$

where $B \in \mathbb{R}^{N_z \times N_z}$ is a given covariance matrix and

$$\text{n}(z; m, B) := \frac{1}{(2\pi)^{N_z/2} |B|^{1/2}} e^{-\frac{1}{2}(z-m)^T B^{-1}(z-m)}.$$

Note that the empirical measure (2.59) is recovered in the limit $B \to 0$. While the weighted empirical measure

$$\pi_{\text{emp}}^{a}(z) := \sum_{i=1}^{M} w_i \delta(z - z_i^f)$$

with weights given by (2.35) provides the analysis in case of an empirical forecast measure (2.59) and an observation y_{obs}, a Gaussian mixture forecast PDF (2.60) leads to an analysis PDF in form of a weighted Gaussian mixture provided the forward operator $h(z)$ is linear in z. This fact allows one to extend the ETPF to Gaussian mixtures. See Reich and Cotter (2015) for more details. Alternative implementations of Gaussian mixture filters can, for example, be found in Stordal et al. (2011) and Frei and Künsch (2013).

Second, one may factorize the likelihood function $\pi_{Y^{1:N}}(y^{1:N}|z^{0,N})$ into $L > 1$ identical copies

$$\hat{\pi}_{Y^{1:N}}(y^{1:N}|z^{0:N}) := \frac{1}{(2\pi)^{N_y N/2}|R/L|^{N/2}} \prod_{n=1}^{N} e^{-\frac{1}{2L}(h(z^n)-y^n)^T R^{-1}(h(z^n)-y^n)},$$

i.e.,

$$\pi_{Y^{1:N}}(y^{1:N}|z^{0,N}) = \prod_{l=1}^{L} \hat{\pi}_{Y^{1:N}}(y^{1:N}|z^{0:N})$$

and one obtains a sequence of L "incremental" Feynman-Kac formulae. Each of these formulae can be approximated numerically by any of the methods discussed in this review. For example, one obtains the continuous EnKF formulation of Bergemann and Reich (2010) in the limit $L \to \infty$ in case of an ESRF. We also mention the continuous Gaussian mixture ensemble transform filter (Reich, 2012). An important advantage of an incremental approach is the fact that the associated weights (2.35) remain closer to the uniform reference value $1/M$ in each iteration step. See also related methods such as *running in place* (RIP) (Kalnay and Yang, 2010), the iterative EnKF approach of Bocquet and Sakov (2012); Sakov et al. (2012), and the embedding approach of Beskov et al. (2013) for SMCMs.

Third, while this paper has been focused on discrete time algorithms, most of the presented results can be extended to differential equations with observations arriving continuously in time such as

$$dy_{obs}(t) = h(z_{ref}(t))dt + \sigma dW(t),$$

where $W(t)$ denotes standard Brownian motion and $\sigma > 0$ determines the amplitude of the measurement error. The associated marginal densities satisfy the Kushner-Stratonovitch stochastic PDE (Jazwinski, 1970). Extensions of the McKean approach to continuous-in-time filtering problems can be found in Crisan and Xiong (2010) and Yang et al. (2013). We also mention the continuous-in-time formulation of the EnKF by Bergemann and Reich (2012). More generally, a reformulation of LETFs in terms of continuously-in-time arriving observations is of the abstract form

$$dz_j = f(z_j)dt + \sum_{i=1}^{M} z_i ds_{ij}, \qquad j = 1, \dots, M. \qquad (2.61)$$

Here $S(t) = \{s_{ij}(t)\}$ denotes a matrix-valued stochastic process which depends on the ensemble $\{z_i(t)\}$ and the observations $y_{obs}(t)$. In other words, (2.61) leads to a particular class of interacting particle systems and we leave further investigations of its properties for future research. We only mention that the continuous-in-time EnKF formulation of Bergemann and Reich (2012) leads to

$$\mathrm{d}s_{ij} = \frac{1}{M-1}(y_i - \bar{y})\sigma^{-1}(\mathrm{d}y_{obs} - y_j\mathrm{d}t + \sigma^{1/2}\mathrm{d}W_j),$$

where the $W_j(t)$'s denote standard Brownian motion, $y_j = h(z_j)$, and $\bar{y} = \frac{1}{M}\sum_{i=1}^{M} y_i$. See also Amezcua et al. (2014) for related reformulations of ESRFs.

Acknowledgements We would like to thank Yann Brenier, Dan Crisan, Mike Cullen, and Andrew Stuart for inspiring discussions on ensemble-based filtering methods and the theory of optimal transportation.

References

J. Amezcua, E. Kalnay, K. Ide, and S. Reich. Ensemble transform Kalman-Bucy filters. *Quarterly J. Royal Meteo. Soc.*, 140:995–1004, 2014.

J.L. Anderson. An adaptive covariance inflation error correction algorithm for ensemble filters. *Tellus*, 59A:210–224, 2007.

J.L. Anderson. A non-Gaussian ensemble filter update for data assimilation. *Monthly Weather Review*, 138:4186–4198, 2010.

J.L. Anderson. Localization and sampling error correction in ensemble Kalman filter data assimilation. *Mon. Wea. Rev.*, 140:2359–2371, 2012.

U.M. Ascher. *Numerical Methods for Evolutionary Differential Equations*. SIAM, Philadelphia, MA, 2008.

A. Bain and D. Crisan. *Fundamentals of Stochastic Filtering*, volume 60 of *Stochastic modelling and applied probability*. Springer-Verlag, New-York, 2009.

T. Bengtsson, P. Bickel, and B. Li. Curse of dimensionality revisited: Collapse of the particle filter in very large scale systems. In *IMS Lecture Notes - Monograph Series in Probability and Statistics: Essays in Honor of David F. Freedman*, volume 2, pages 316–334. Institute of Mathematical Sciences, 2008.

K. Bergemann and S. Reich. A localization technique for ensemble Kalman filters. *Q. J. R. Meteorological Soc.*, 136:701–707, 2010.

K. Bergemann and S. Reich. An ensemble Kalman-Bucy filter for continuous data assimilation. *Meteorolog. Zeitschrift*, 21:213–219, 2012.

A. Beskov, D. Crisan, A. Jasra, and N. Whiteley. Error bounds and normalizing constants in sequential Monte Carlo in high dimensions. *Adv. App. Probab.*, 2013. to appear.

M. Bocquet and P. Sakov. Combining inflation-free and iterative ensemble Kalman filters for strongly nonlinear systems. *Nonlin. Processes Geophys.*, 19:383–399, 2012.

M. Bonavita, L. Isaksen, and E. Holm. On the use of EDA background error variances in the ECMWF 4D-Var. *Q. J. Royal. Meteo. Soc.*, 138:1540–1559, 2012.

G. Burgers, P.J. van Leeuwen, and G. Evensen. On the analysis scheme in the ensemble Kalman filter. *Mon. Wea. Rev.*, 126:1719–1724, 1998.

R.E. Caflisch. Monte Carlo and quasi-Monte Carlo methods. In *Acta Numerica*, volume 7, pages 1–49. Cambridge University Press, 1988.

R.E. Caflisch, W. Morokoff, and A.B. Owen. Valuation of mortgage backed securities using Brownian bridges to reduce effective dimension. *Journal of Computational Finance*, pages 27–46, 1997.

A.J. Chorin, M. Morzfeld, and X. Tu. Implicit filters for data assimilation. *Comm. Appl. Math. Comp. Sc.*, 5:221–240, 2010.

A.M. Clayton, A.C. Lorenc, and D.M. Barker. Operational implementation of a hybrid ensemble/4D-Var global data assimilation system at the Met Office. *Q. J. Royal. Meteo. Soc.*, 139:1445–1461, 2013.

S.L. Cotter, M. Dashti, J.C. Robinson, and A.M. Stuart. Bayesian inverse problems for functions and applications to fluid mechanics. *Inverse Problems*, 25:115008, 2009.

D. Crisan and J. Xiong. Approximate McKean-Vlasov representation for a class of SPDEs. *Stochastics*, 82:53–68, 2010.

M. Cuturi. Sinkhorn distances: Lightspeed computation of optimal transport. In *Advances in Neural Information Processing Systems*, pages 2292–2300, Lake Tahoe, Nevada, 2013.

R. Daley. *Atmospheric Data Analysis*. Cambridge University Press, Cambridge, 1993.

M. Dashti, K.J.H. Law, A.M. Stuart, and J. Voss. MAP estimators and posterior consistency. *Inverse Problems*, 29:095017, 2013.

P. del Moral. *Feynman-Kac Formulae: Genealogical and Interacting Particle Systems with Applications*. Springer-Verlag, New York, 2004.

A. Doucet and A.M. Johansen. A tutorial on particle filtering and smoothing: fifteen years later. In D. Crisan and B. Rozovskii, editors, *Oxford Handbook of Nonlinear Filtering*, pages 656–704, 2011.

A. Doucet, N. de Freitas, and N. Gordon (eds.). *Sequential Monte Carlo Methods in Practice*. Springer-Verlag, Berlin Heidelberg New York, 2001.

G. Evensen. Sequential data assimilation with a nonlinear quasigeostrophic model using Monte Carlo methods for forecasting error statistics. *J. Geophys. Res.*, 99:10143–10162, 1994.

G. Evensen. *Data Assimilation. The Ensemble Kalman Filter*. Springer-Verlag, New York, 2006.

M. Frei and H.R. Künsch. Mixture ensemble Kalman filters. *Computational Statistics and Data Analysis*, 58:127–138, 2013.

G. Gaspari and S.E. Cohn. Construction of correlation functions in two and three dimensions. *Q. J. Royal Meteorological Soc.*, 125:723–757, 1999.

G.H. Golub and Ch.F. van Loan. *Matrix Computations*. The Johns Hopkins University Press, Baltimore, 3rd edition, 1996.

C. González-Tokman and B.R. Hunt. Ensemble data assimilation for hyperbolic systems. *Physica D*, 243:128–142, 2013.

N.J. Gordon, D.J. Salmon, and A.F.M. Smith. Novel approach to nonlinear/non-Gaussian Bayesian state estimation. *IEEE Proceedings F on Radar and Signal Processing*, 140:107–113, 1993.

Th.M. Hamill, J.S. Whitaker, and Ch. Snyder. Distance-dependent filtering of background covariance estimates in an ensemble Kalman filter. *Mon. Wea. Rev.*, 129:2776–2790, 2001.

J.E. Handschin and D.Q. Mayne. Monte Carlo techniques to estimate the conditional expectation in multi-stage non-linear filtering. *Internat. J. Control*, 1:547–559, 1969.

P.L. Houtekamer and H.L. Mitchell. A sequential ensemble Kalman filter for atmospheric data assimilation. *Mon. Wea. Rev.*, 129:123–136, 2001.

P.L. Houtekamer and H.L. Mitchell. Ensemble Kalman filtering. *Q. J. Royal Meteorological Soc.*, 131:3269–3289, 2005.

B.R. Hunt, E.J. Kostelich, and I. Szunyogh. Efficient data assimilation for spatialtemporal chaos: A local ensemble transform Kalman filter. *Physica D*, 230:112–137, 2007.

T. Janjić, D. McLaughlin, S.E. Cohn, and M. Verlaan. Conservation of mass and preservation of positivity with ensemble-type Kalman filter algorithms. *Mon. Wea. Rev.*, 142:755–773, 2014.

A.H. Jazwinski. *Stochastic Processes and Filtering Theory*. Academic Press, New York, 1970.

J. Kaipio and E. Somersalo. *Statistical and Computational Inverse Problems*. Springer-Verlag, New York, 2005.

E. Kalnay. *Atmospheric Modeling, Data Assimilation and Predictability*. Cambridge University Press, 2002.

E. Kalnay and S.-C. Yang. Accelerating the spin-up of ensemble Kalman filtering. *Quart. J. Roy. Meteor. Soc.*, 136:1644–1651, 2010.

P.K. Kitanidis. Quasi-linear geostatistical theory for inverting. *Water Resources Research*, 31: 2411–2419, 1995.

K.J.H. Law, A. Shukia, and A.M. Stuart. Analysis of the 3DVAR filter for the partially observed Lorenz'63 model. *Discrete and Continuous Dynamical Systems A*, 2013. to appear.

T.S. Lee and D. Mumford. Hierarchical Bayesian inference in the visual cortex. *J. Opt. Soc. Am. A*, 20:1434–1448, 2003.

J. Lei and P. Bickel. A moment matching ensemble filter for nonlinear and non-Gaussian data assimilation. *Mon. Weath. Rev.*, 139:3964–3973, 2011.

C.E. Leith. Theoretical skills of Monte Carlo forecasts. *Mon. Weath. Rev.*, 102:409–418, 1974.

T. Lelièvre, M. Rousset, and G. Stoltz. *Free Energy Computations - A Mathematical Perspective*. Imperial College Press, London, 2010.

J.M Lewis, S. Lakshmivarahan, and S.K. Dhall. *Dynamic Data Assimilation: A Least Squares Approach*. Cambridge University Press, Cambridge, 2006.

J.S. Liu. *Monte Carlo Strategies in Scientific Computing*. Springer-Verlag, New York, 2001.

D.M. Livings, S.L. Dance, and N.K. Nichols. Unbiased ensemble square root filters. *Physica D*, 237:1021–1028, 2008.

E.N. Lorenz. Deterministic non-periodic flows. *J. Atmos. Sci.*, 20:130–141, 1963.

E.N. Lorenz. Predictability: A problem partly solved. In *Proc. Seminar on Predictability*, volume 1, pages 1–18, ECMWF, Reading, Berkshire, UK, 1996.

A. Majda and J. Harlim. *Filtering Complex Turbulent Systems*. Cambridge University Press, Cambridge, 2012.

R.J. McCann. Existence and uniqueness of monotone measure-preserving maps. *Duke Mathematical Journal*, 80:309–323, 1995.

H.P. McKean. A class of Markov processes associated with nonlinear parabolic equations. *Proc. Natl. Acad. Sci. USA*, 56:1907–1911, 1966.

S. Metref, E. Cosme, C. Snyder, and P. Brasseur. A non-Gaussian analysis scheme using rank histograms for ensemble data assimilation. *Nonlinear Processes in Geophysics*, 21:869–885, 2013.

T. Miyoshi. The Gaussian approach to adaptive covariance inflation and its implementation with the local ensemble transform Kalman filter. *Mon. Wea. Rev.*, 139:1519–1535, 2011.

K.W. Morton and D.F. Mayers. *Numerical Solution of Partial Differential Equations*. Cambridge University Press, Cambridge, 2nd edition, 2005.

M. Morzfeld and A.J. Chorin. Implicit particle filtering for models with partial noise and an application to geomagnetic data assimilation. *Nonlinear Processes in Geophysics*, 19:365–382, 2012.

M. Morzfeld, X. Tu, E. Atkins, and A.J. Chorin. A random map implementation of implicit filters. *J. Comput. Phys.*, 231:2049–2066, 2012.

T.A. El Moselhy and Y.M. Marzouk. Bayesian inference with optimal maps. *J. Comput. Phys.*, 231:7815–7850, 2012.

L. Nerger, T. Janijc Pfander, J. Schröter, and W. Hiller. A regulated localization scheme for ensemble-based Kalman filters. *Quarterly J. Royal Meteo. Soc.*, 138:802–812, 2012.

J. Nocedal and S.J. Wright. *Numerical Optimization*. Springer-Verlag, New York, 2nd edition, 2006.

D.S. Oliver. On conditional simulation to inaccurate data. *Math. Geology*, 28:811–817, 1996.

D.S. Oliver, N. He, and A.C. Reynolds. Conditioning permeability fields on pressure data. Technical report, presented at the 5th European Conference on the Mathematics of Oil Recovery, Leoben, Austria, 1996.

I. Olkin and F. Pukelsheim. The distance between two random vectors with given dispersion matrices. *Linear Algebra and its Applications*, 48:257–263, 1982.

E. Ott, B.R. Hunt, I. Szunyogh, A.V. Zimin, E.J. Kostelich, M. Corazza, E. Kalnay, D.J. Patil, and J.A. Yorke. A local ensemble Kalman filter for atmospheric data assimilation. *Tellus*, A 56: 415–428, 2004.

O. Pele and M. Werman. Fast and robust earth mover's distances. In *Computer Vision, 2009 IEEE 12th international conference*, pages 460–467, 2009.

D.T. Pham. Stochastic methods for sequential data assimilation in strongly nonlinear systems. *Mon. Wea. Rev.*, 129:1194–1207, 2001.

P. Rebeschini and R. van Handel. Can local particle filters beat the curse of dimensionality? *arXiv:1301.6585*, 2013.

S. Reich. A dynamical systems framework for intermittent data assimilation. *BIT Numer Math*, 51: 235–249, 2011.

S. Reich. A Gaussian mixture ensemble transform filter. *Q. J. R. Meterolog. Soc.*, 138:222–233, 2012.

S. Reich. A nonparametric ensemble transform method for Bayesian inference. *SIAM J. Sci. Comput.*, 35:A2013–A2024, 2013a.

S. Reich. A guided sequential Monte Carlo method for the assimilation of data into stochastic dynamical systems. In *Recent Trends in Dynamical Systems*, volume 35 of *Springer Proceedings in Mathematics and Statistics*, pages 205–220, 2013b.

S. Reich and C. J. Cotter. Ensemble filter techniques for intermittent data assimilation. In M. Cullen, Freitag M. A., S. Kindermann, and R. Scheichl, editors, *Large Scale Inverse Problems. Computational Methods and Applications in the Earth Sciences*, volume 13 of *Radon Ser. Comput. Appl. Math.*, pages 91–134. Walter de Gruyter, Berlin, 2013.

S. Reich and C.J. Cotter. *Probabilistic Forecasting and Bayesian Data Assimilation*. Cambridge University Press, Cambridge, 2015.

L.F. Richardson. *Weather Prediction by Numerical Processes*. Cambridge University Press, Cambridge, 1922.

Ch. Robert and G. Casella. *Monte Carlo Statistical Methods*. Springer-Verlag, New York, 2nd edition, 2004.

J.C. Robinson. *Infinite-Dimensional Dynamical Systems*. Cambridge University Press, Cambridge, 2001.

P. Sakov, D. Oliver, and L. Bertino. An iterative EnKF for strongly nonlinear systems. *Mon. Wea. Rev.*, 140:1988–2004, 2012.

S. Särkkä. *Bayesian Filtering and Smoothing*. Cambridge University Press, Cambridge, 2013.

A.S. Stordal, H.A. Karlsen, G. Nævdal, H.J. Skaug, and B. Vallés. Bridging the ensemble Kalman filter and particle filters: The adaptive Gaussian mixture filter. *Comput. Geosci.*, 15:293–305, 2011.

G. Strang. *Introduction to Applied Mathematics*. Wellesley-Cambridge Press, 2nd edition, 1986.

A.M. Stuart. Inverse problems: a Bayesian perspective. In *Acta Numerica*, volume 17, pages 451–559. Cambridge University Press, Cambridge, 2010.

A. Tarantola. *Inverse Problem Theory and Methods for Model Parameter Estimation*. SIAM, Philadelphia, 2005.

M.K. Tippett, J.L. Anderson, G.H. Bishop, T.M. Hamill, and J.S. Whitaker. Ensemble square root filters. *Mon. Wea. Rev.*, 131:1485–1490, 2003.

P.J. van Leeuwen. Nonlinear data assimilation in the geosciences: An extremely efficient particle filter. *Q.J.R. Meteorolog. Soc.*, 136:1991–1996, 2010.

P.J. van Leeuwen and M. Ades. Efficient fully nonlinear data assimilation for geophysical fluid dynamics. *Computers and Geosciences*, 47:in press, 2012.

C. Villani. *Topics in Optimal Transportation*. American Mathematical Society, Providence, Rhode Island, NY, 2003.

C. Villani. *Optimal Transportation: Old and New*. Springer-Verlag, Berlin Heidelberg, 2009.

X. Wang, C.H. Bishop, and S.J. Julier. Which is better, an ensemble of positive-negative pairs or a centered spherical simplex ensemble? *Mon. Wea. Rev.*, 132:1590–1505, 2004.

N. Wiener. *Extrapolation, Interpolation and Smoothing of Stationary Time Series: With Engineering Applications*. MIT Press, Cambridge, MA, 1948.

T. Yang, P.G. Mehta, and S.P. Meyn. Feedback particle filter. *IEEE Trans. Automatic Control*, 58: 2465–2480, 2013.